An Introduction to the FUNCTIONAL FORMULATION of QUANTUM MECHANICS

Horacio Oscar Girotti
Universidade Federal do Rio Grande do Sul, Brazil

An Introduction to the FUNCTIONAL FORMULATION of QUANTUM MECHANICS

NEW JERSEY • LONDON • SINGAPORE • BEIJING • SHANGHAI • HONG KONG • TAIPEI • CHENNAI

Published by

World Scientific Publishing Co. Pte. Ltd.
5 Toh Tuck Link, Singapore 596224
USA office: 27 Warren Street, Suite 401-402, Hackensack, NJ 07601
UK office: 57 Shelton Street, Covent Garden, London WC2H 9HE

Library of Congress Cataloging-in-Publication Data
Girotti, Horacio Oscar.
An introduction to the functional formulation of quantum mechanics / by Horacio Oscar Girotti, Universidade Federal do Rio Grande do Sul, Brazil.
pages cm
Includes bibliographical references and index.
ISBN 978-981-4520-49-2 (hardcover : alk. paper)
1. Quantum theory. 2. Functionals. I. Titl
QC174.17.F86G57 2013
530.12--dc23

2013018807

British Library Cataloguing-in-Publication Data
A catalogue record for this book is available from the British Library.

Printed in Singapore

To Mabel

Preface

The functional formulation of quantum mechanics and relativistic quantum field theories are widely studied subjects. Nevertheless, certain aspects of this formulation are frequently overlooked, in spite of their relevance from a structural point of view. This book aims at filling up this gap. The non-relativistic regime has been found particularly appropriate for these purposes.

In Chapter 1 we are mainly concerned with the lack of uniqueness pervading the transition from the operator to the functional formulation of quantum mechanics. As is known, the propagator for a finite time interval is the object of primary interest in the functional framework; it is to be written in terms of a phase space path integral. In order to reach this goal we can start by slicing the time interval into subintervals of equal size. Thereby, it will be possible to express the propagator as a product of amplitudes commonly referred to as short-time propagators. The Generalized Weyl Transform, a correspondence rule associating operators with c-number functions and vice versa, is used to map the evolution operator entering each short time propagator into a set of classical functions. This process gives rise, after completion, to an infinite set of phase space path integrals. There is no proof that all integrals in this set yield the same answer. In the standard formulation of the harmonic oscillator, for instance, uniqueness follows straightforwardly. However, after a reformulation of the dynamics of this system via a canonical transformation the proof of uniqueness calls for a rather intricate mechanism based on the stochastic nature of the path integral.

Chapter 2 deals with a unified view of the functional formulation of quantum mechanics. Here we discuss the functional differential scheme put forward by Schwinger and its equivalence to the phase space path integral formulation. The fact that the fictitious sources linearly enter the Schwinger equations is at the root of this result. Indeed, this linearity suggests the functional Fourier transform as a natural tool for solving Schwinger equations. It is amusing that this Fourier transform turns out to be the phase space path integral. The interplay between the Generalized Weyl Transform and Schwinger equations is also analyzed.

In Chapter 3 the focus is on computing the connected Green functions generating functional. We first concentrate on the one dimensional harmonic oscillator. This system serves as an arena for implementing a systematics of general validity. The connection between the discontinuities of the two-point Green functions and the canonical commutation relations is established. Systems whose energy eigenstates are non-normalizable are also analyzed. We then address the problem of computing the connected Green functions generating functional in full generality. Accordingly, we use the perturbative technique known as loop expansion. The one loop quantum corrections for the effective action and the effective potential are determined for a specific model.

In Chapter 4 we review the Hamiltonian formulation of the classical dynamics of gauge systems as well as their functional quantization.

The quantum dynamics of systems involving noncommutative coordinates, to be referred to as noncommutative systems, is presented in Chapter 5. Our purpose here is to expose the reader to a recent field of research. In truth, the use of noncommutative space-time coordinates in field theory was proposed by Heisenberg to circumvent the problem raised by the ultraviolet divergences. The idea was override by the success of renormalization theory. Its revival is rather recent and in connection with the low energy limit in superstrings. The attractive feature is that noncommutative systems can be thought as constrained systems possessing second class constraints. Our developments in Chapter 4 serve, then, for the purpose of quantizing noncommutative systems.

A useful set of integrals along with some preliminary notions on functional analysis have been relegated to the appendices.

I am indebted to Professor Miguel Ángelo Cavalheiro Gusmão (IF-UFRGS) for discussions. Support from Conselho Nacional de Desenvolvimento Científico e Tecnológico (CNPq), Brazil, is also acknowledged.

H. O. Girotti

Contents

Chapter 1

Correspondence rules. The phase space path integral

The opening section summarizes the highlights of the operator formulation of quantum mechanics. The specification of the physical system and the definition of the propagator are also included here. Furthermore in this chapter, we define and analyze the generalized Weyl transform (GWT). Of particular interest is the link relating the GWT with the ordering of noncommuting operators. The time-slicing procedure and the GWT are used to build up the phase space path integral representative for the propagator. In fact, the just mentioned amplitude turns out to be associated with an infinite set of phase space path integrals, each element in the set being labeled by a real parameter. The uniqueness of the formulation calls for the mutual cancellation of all terms depending on this parameter. This holds true for systems whose Hamiltonian operators do not contain products of noncommuting operators. Otherwise, an explicit example suggests that the desired cancellation might take place as a by-product of the stochastic nature of the phase space path integral. The one dimensional harmonic oscillator as well as the one dimensional free particle and the modified one dimensional harmonic oscillator are worked out in detail.

1.1 Operator formulation of quantum mechanics

Here, the quantities belonging to the Schrödinger and Heisenberg pictures will be denoted by the subscripts or superscripts S and H, respectively.

1.1.1 *Schrödinger picture*

The Schrödinger picture state vector $|\psi(t)\rangle_S$ evolves in time in accordance with the Schrödinger equation

$$H_S(t)\,|\psi(t)\rangle_S \;=\; i\,\hbar\,\frac{d|\psi(t)\rangle_S}{dt}\,, \tag{1.1}$$

while operators $\{\mathcal{O}_S\}$ are required to satisfy

$$\frac{d\mathcal{O}_S(t)}{dt} \;=\; \frac{\partial\mathcal{O}_S(t)}{\partial t}\,. \tag{1.2}$$

Observables are represented by self-adjoint operators [1].

The change in time of the state vector can be phrased in terms of the *time evolution operator* $U_S(t,t_i)$. It is defined through the mapping

$$|\psi(t_i)\rangle_S \longrightarrow |\psi(t)\rangle_S \;=\; U_S(t,t_i)\,|\psi(t_i)\rangle_S\,, \quad t \ge t_i\,, \tag{1.3}$$

where t_i is some initial time. The operator $U_S(t,t_i)$ fulfils the differential equation

$$H_S(t)\,U_S(t,t_i) \;=\; i\,\hbar\,\frac{dU_S(t,t_i)}{dt}\,, \tag{1.4}$$

the initial condition

$$U_S(t_i,t_i) \;=\; I\,, \tag{1.5}$$

and the group composition law

$$U_S(t,t_i) \;=\; U_S(t,t')\,U_S(t',t_i)\,, \qquad t \ge t' \ge t_i\,. \tag{1.6}$$

One can verify that Eq.(1.4) defines a *unitary operator*. Hence, the dynamics does not alter the norm of the state vector and, accordingly, we are allowed to choose

$${}_S\langle\psi(t)|\psi(t)\rangle_S \;=\; 1\,, \qquad \forall\, t \ge t_i\,. \tag{1.7}$$

The integral equation

$$U_S(t,t_i) \;=\; I \;-\; \frac{i}{\hbar}\int_{t_i}^{t} dt'\,H_S(t')\,U_S(t',t_i)\,, \tag{1.8}$$

[1] The Hermitian conjugate of the operator $\mathcal{O}$ will be denoted by $\mathcal{O}^\dagger$. Self-adjoint operators verify: i) $\mathcal{O} = \mathcal{O}^\dagger$ and ii) $\mathcal{D}(\mathcal{O}) = \mathcal{D}(\mathcal{O}^\dagger)$. Here, $\mathcal{D}(\mathcal{O})$ is the domain of definition of $\mathcal{O}$.

summarizes the differential equation (1.4) and the initial condition (1.5). The ordering of operators on the right hand side of Eq.(1.8) matters unless H_S is time independent. In this case, Eq.(1.8) can be explicitly integrated to yield

$$U_S(t, t_i) = \exp\left[-\frac{i}{\hbar} H_S(t - t_i)\right] . \tag{1.9}$$

From now on we shall restrict ourselves to deal with physical systems whose dynamics is specified by time independent Hamiltonian operators.

Let now $\mathcal{S}_S$ be one of the maximal sets of commuting observables containing the Hamiltonian H_S as one of its elements. The observables in $\mathcal{S}_S$ possess a unique set of common eigenvectors $\{|E_k, \beta\rangle_S\}$ which provides a basis in the space of states [2]. Therefore, we have that

$$H_S \, |E_k, \beta\rangle_S = E_k \, |E_k, \beta\rangle_S \, , \tag{1.10}$$

$${}_S\langle E_k, \beta | E_{k'}, \beta'\rangle_S = \delta_{kk'} \, \delta_{\beta\beta'} \, , \tag{1.11}$$

and [3]

$$I = \sum_{k,\beta} |E_k, \beta\rangle_S \, {}_S\langle E_k, \beta| \, . \tag{1.12}$$

Therefore, the state vector at the initial time can be written as

$$|\psi(t_i)\rangle_S = \sum_{k,\beta} |E_k, \beta\rangle_S \, {}_S\langle E_k, \beta|\psi(t_i)\rangle_S = \sum_{k,\beta} c^{(S)}_{k,\beta}(t_i) \, |E_k, \beta\rangle_S \, , \tag{1.13}$$

where the amplitudes

$$c^{(S)}_{k,\beta}(t_i) \equiv_S \langle E_k, \beta|\psi(t_i)\rangle_S \tag{1.14}$$

verify

$$\sum_{k,\beta} |c^{(S)}_{k,\beta}(t_i)|^2 = 1 \, , \tag{1.15}$$

---

[2] We have isolated the energy eigenvalue and grouped the remaining ones into the single label β.

[3] The summation symbol implies sum over the discrete indices and integration over the continuous ones. The delta symbols on the right hand side of Eq.(1.11) must be understood respectively as Kroenecker or Dirac delta functions depending on whether the labels attached to them are discrete or continuous, respectively.

in order to comply with the normalization condition in Eq.(1.7). By combining Eqs.(1.13), (1.3), (1.9) and (1.10) we obtain

$$|\psi(t)\rangle_S = \sum_{k,\beta} c_{k,\beta}^{(S)}(t)\,|E_k,\beta\rangle_S\,, \tag{1.16}$$

where

$$c_{k,\beta}^{(S)}(t) = \exp\left[-\frac{i}{\hbar}\,E_k\,(t-t_i)\right]\,c_{k,\beta}^{(S)}(t_i) \tag{1.17}$$

is the amplitude for finding the system in the state $|E_k,\beta\rangle_S$ at time t.

1.1.2 *Heisenberg picture*

Without restrictions on H_S, the Heisenberg and the Schrödinger pictures are related by way of the canonical transformation

$$|\psi(t)\rangle_H = U_S^\dagger(t,t_i)\,|\psi(t)\rangle_S\,, \tag{1.18a}$$
$$\mathcal{O}_H(t) = U_S^\dagger(t,t_i)\,\mathcal{O}_S(t)\,U_S(t,t_i)\,. \tag{1.18b}$$

Clearly, we have chosen $t = t_i$ as the instant of time at which both pictures coincide.

The dynamics in the Heisenberg picture derives from that in the Schrödinger picture by taking advantage of Eqs.(1.18). We can check that

$$\frac{d|\psi(t)\rangle_H}{dt} = 0 \tag{1.19}$$

while

$$\frac{d\mathcal{O}_H(t)}{dt} = \frac{i}{\hbar}\,[H_H(t)\,,\,\mathcal{O}_H(t)] + \frac{\partial\mathcal{O}_H(t)}{\partial t}\,. \tag{1.20}$$

Hence, in the Heisenberg picture the dynamics is fully carried out by the operators. As for the Hamiltonian, Eq.(1.20) reduces to

$$\frac{dH_H(t)}{dt} = \frac{\partial H_H(t)}{\partial t}\,. \tag{1.21}$$

Let $\mathcal{S}_H$ be the Heisenberg picture image of $\mathcal{S}_S$. Each pair of observables connected by a canonical transformation share a common set of real eigenvalues. According to Eq.(1.18a), the eigenvectors are in turn correlated through

$$|E_k,\beta;t\rangle_H = U_S^\dagger(t,t_i)\,|E_k,\beta\rangle_S\,, \tag{1.22}$$

which for a time independent Hamiltonian operator becomes

$$|E_k, \beta; t\rangle_H = \exp\left[\frac{i}{\hbar} E_k(t - t_i)\right] |E_k, \beta\rangle_S . \tag{1.23}$$

Moreover, the spectral resolution of the identity operator (see Eq.(1.12)) written in terms of Heisenberg picture eigenvectors reads

$$I = \sum_{k,\beta} |E_k, \beta; t\rangle_H \, {}_H\langle E_k, \beta; t| . \tag{1.24}$$

Concerning the stationary Heisenberg picture state vector, it can be written as the linear superposition of eigenvectors

$$|\psi\rangle_H = \sum_{k,\beta} |E_k, \beta; t\rangle_H \, {}_H\langle E_k, \beta; t|\psi\rangle_H = \sum_{k,\beta} c^{(H)}_{k,\beta}(t) \, |E_k, \beta; t\rangle_H , \tag{1.25}$$

where

$$\begin{aligned} c^{(H)}_{k,\beta}(t) &= {}_H\langle E_k, \beta; t|\psi\rangle_H = \exp\left[-\frac{i}{\hbar} E_k(t - t_i)\right] \, {}_S\langle E_k, \beta|\psi\rangle_H \\ &= \exp\left[-\frac{i}{\hbar} E_k(t - t_i)\right] \, {}_S\langle E_k, \beta|\psi(t_i)\rangle_S \\ &= \exp\left[-\frac{i}{\hbar} E_k(t - t_i)\right] \, c^{(S)}_{k,\beta}(t_i) = c^{(S)}_{k,\beta}(t) . \end{aligned} \tag{1.26}$$

This confirms that probability amplitudes are picture independent.

1.1.3 *The physical system*

We shall be considering a physical system whose phase space Γ is spanned by the linearly independent bosonic variables $q_i, p_i, i = 1, 2, \ldots, N$, where $\{q_i | \, i = 1, 2, \ldots, N\}$ are Cartesian coordinates and $\{p_i | \, i = 1, 2, \ldots, N\}$ their corresponding canonical conjugate momenta. The classical dynamics is generated by the Hamiltonian $h(q, p)$, which is assumed to be a real function possessing a lower bound in Γ. The Schrödinger picture quantum counterparts of q_i, p_i and $h(q, p)$ will be denoted by the capital letters Q_i, P_i and $H(Q, P)$, respectively. The picture label S will be suppressed since the quantities, vectors and operators, that we shall be dealing with in the remaining sections of this chapter belong mostly to the Schrödinger picture.

In accordance with the correspondence principle, the system is to be quantized by abstracting the basic commutators from the corresponding Poisson brackets. This gives rise to the canonical algebra

$$[Q_i\,,\,Q_j] = 0\,, \tag{1.27a}$$
$$[Q_i\,,\,P_j] = i\,\hbar\,\delta_{ij}\,I\,, \tag{1.27b}$$
$$[P_i\,,\,P_j] = 0\,, \tag{1.27c}$$

where $\hbar$ is the Dirac constant. The transition $h(q,p) \longrightarrow H(Q,P)$ may be afflicted by ordering ambiguities.

We look next for a representation of the algebra in Eq.(1.27). This implies finding a basis in the space of states which will allow for realizing the operators in terms of matrices. The set of the common eigenvectors

$$|q\rangle = |q_1, q_2, \ldots, q_N\rangle \equiv |q_1\rangle \otimes |q_2\rangle \otimes \ldots |q_N\rangle \tag{1.28}$$

of the position observables $Q_i, i = 1, \ldots, N$ provides such basis. They solve the eigenvalue problem

$$Q_i\,|q\rangle = q_i\,|q\rangle\,, i = 1, 2, , \ldots, N. \tag{1.29}$$

Since we are dealing with Cartesian coordinates, the eigenvalues $\{q_i, i = 1, \ldots, N\}$ run continuously from $-\infty$ to $+\infty$. Correspondingly, the eigenvectors are normalized as

$$\langle q|q'\rangle = \delta^{(N)}(q - q')\,, \tag{1.30}$$

where

$$\delta^{(N)}(q - q') = \int_{-\infty}^{+\infty} \frac{d^N p}{(2\pi\hbar)^N}\, e^{\frac{i}{\hbar}\,p\cdot(q-q')}\,, \tag{1.31}$$

$$d^N p \equiv \prod_{i-1}^{N} dp_i\,, \tag{1.32}$$

and [4]

$$p \cdot q \equiv p_i\,q_i\,. \tag{1.33}$$

[4] In a monomial, the sum from 1 to N over repeated indices is always implied.

The spectral resolution of the identity

$$I = \int_{-\infty}^{+\infty} d^N q \, |q\rangle\langle q| \,, \tag{1.34}$$

is consistent with Eq.(1.30). This secures the completeness of the set $\{|q\rangle\}$. As is known, the (continuous) matrices representing the operators Q_i and P_i in the coordinate basis $\{|q\rangle\}$ are, respectively,

$$\langle q|Q_i|q'\rangle = q_i \, \delta^{(N)}(q - q') \,, \tag{1.35a}$$

$$\langle q|P_i|q'\rangle = -\, i\, \hbar \frac{\partial}{\partial q_i} \delta^{(N)}(q - q') \,. \tag{1.35b}$$

Solving the linear momentum eigenvalue problem

$$P_i \, |p\rangle = p_i \, |p\rangle \,, \tag{1.36}$$

where

$$|p\rangle = |p_1, p_2, \ldots, p_N\rangle \equiv |p_1\rangle \otimes |p_2\rangle \otimes \ldots |p_N\rangle \,, \tag{1.37}$$

is also of interest for our purposes. The position representation of (1.36) is

$$-\, i\, \hbar \frac{\partial \langle q|p\rangle}{\partial q_i} = p_i \, \langle q|p\rangle \tag{1.38}$$

whose solution reads

$$\langle q|p\rangle = C \, e^{\frac{i}{\hbar} q\cdot p} \,, \tag{1.39}$$

with the eigenvalues $\{p_i\}$ running continuously from $-\infty$ to $+\infty$. The integration constant C is found by demanding

$$\langle p|p'\rangle = |C|^2 \int_{-\infty}^{+\infty} d^N q \, e^{-\frac{i}{\hbar} q\cdot(p - p')} = \delta^{(N)}(p - p') \,. \tag{1.40}$$

We end up with

$$\langle q|p\rangle = \frac{1}{(2\pi\hbar)^{\frac{N}{2}}} \, e^{\frac{i}{\hbar} q\cdot p} \,. \tag{1.41}$$

The set of linear momentum eigenvectors $\{|p\rangle\}$ is complete and provides another basis in the space of states. Indeed,

$$\int_{-\infty}^{+\infty} d^N p \langle q|p\rangle\langle p|q'\rangle = \frac{1}{(2\pi\hbar)^N}\int_{-\infty}^{+\infty} d^N p \, e^{\frac{i}{\hbar}(q-q')\cdot p}$$
$$= \delta^{(N)}(q-q') = \langle q|I|q'\rangle \tag{1.42}$$

which in view of the arbitrariness of the states $|q\rangle$ and $|q'\rangle$ implies that

$$I = \int_{-\infty}^{+\infty} d^N p |p\rangle\langle p| \tag{1.43}$$

which corroborates the previous statement.

1.1.4 *The propagator*

Within the functional formulation the quantum mechanical amplitudes are given in terms of path integrals. We shall be looking for the path integral describing the amplitude

$$K(q_f, t_f; q_i, t_i) \equiv \langle q_f| \, e^{-\frac{i}{\hbar} H(Q,P)(t_f - t_i)} \, |q_i\rangle \tag{1.44}$$

which is known as the *propagator.*

The propagator plays a key role. Indeed, from Eq.(1.9) it follows that the coordinate representative of Eq.(1.3),

$$\langle q_f|\psi(t_f)\rangle = \int d^N q_i \, \langle q_f|e^{-\frac{i}{\hbar}H(t_f-t_i)}|q_i\rangle\langle q_i|\psi(t_i)\rangle \,, \tag{1.45}$$

can be cast

$$\psi(q_f, t_f) = \int d^N q_i \, K(q_f, t_f; q_i, t_i) \, \psi(q_i, t_i) \,, \tag{1.46}$$

where $\psi(q_i, t_i) \equiv \langle q_i|\psi(t_i)\rangle$ and $\psi(q_f, t_f) \equiv \langle q_f|\psi(t_f)\rangle$ are the initial and final state wave functions, respectively. Clearly, the kernel $K(q_f, t_f; q_i, t_i)$ *propagates* the system from the initial to the final state. Furthermore, the coordinate representative of the boundary condition (1.5) reads

$$\lim_{t_f \downarrow t_i} K(q_f, t_f; q_i, t_i) = \delta^{(N)}(q_f - q_i) \,. \tag{1.47}$$

In terms of the Heisenberg picture position eigenstates, the propagator can be written as

$$K(q_f, t_f; q_i, t_i) = {}_H\langle q_f, t_f | q_i, t_i\rangle_H \,, \tag{1.48}$$

which can be corroborated by using Eqs.(1.18). Hence, if the stationary Heisenberg picture state vector coincides with the position eigenstate $|q_i, t_i\rangle_H$, the propagator will be the probability amplitude to find the system in $q = q_f$ at $t = t_f$.

1.2 Correspondence rules

1.2.1 *The generalized Weyl transform*

The *Generalized Weyl Transform* is a particular case of the *correspondence rules* formulated by Cohen [Cohen (1966, 1976)] and Agarwal and Wolf [Agarwal and Wolf (1970)]. It maps a classical phase space function $a(q, p)$ into a class of operators $A_\alpha(Q, P)$ and vice versa, i.e.,

$$a(q, p) \xrightarrow{\alpha} A_\alpha(Q, P) \,, \tag{1.49a}$$

$$a_\alpha(q, p) \xleftarrow{\alpha} A(Q, P) \,. \tag{1.49b}$$

In short [Leschke and Schmutz (1977)],

$$A_\alpha(Q, P) \equiv \int_{-\infty}^{+\infty} d^N p \int_{-\infty}^{+\infty} d^N q \, a(q, p) \, \Delta_\alpha(Q - q, P - p) \,, \tag{1.50}$$

where

$$\begin{aligned} \Delta_\alpha(Q - q, P - p) &\equiv (2\pi\hbar)^{-N} \\ \times \int_{-\infty}^{+\infty} d^N \tau \, e^{-\frac{i}{\hbar}\tau\cdot p} &\left| q - \left(\frac{1}{2} + \alpha\right)\tau \right\rangle \left\langle q + \left(\frac{1}{2} - \alpha\right)\tau \right| = (2\pi\hbar)^{-2N} \\ \times \int_{-\infty}^{+\infty} d^N u \int_{-\infty}^{+\infty} d^N v \, & e^{\frac{i}{\hbar}[(q-Q)\cdot u + (p-P)\cdot v]} \, e^{\frac{i}{\hbar}\alpha v\cdot u} \,, \end{aligned} \tag{1.51}$$

and

$$\begin{aligned} a_\alpha(q, p) \equiv \int_{-\infty}^{+\infty} d^N \tau \, e^{\frac{i}{\hbar}\tau\cdot p} \\ \times \left\langle q - \left(\frac{1}{2} - \alpha\right)\tau \right| A(Q, P) \left| q + \left(\frac{1}{2} + \alpha\right)\tau \right\rangle . \end{aligned} \tag{1.52}$$

Here, α is a real parameter such that

$$-\frac{1}{2} \leq \alpha \leq +\frac{1}{2} \,. \tag{1.53}$$

For $\alpha = 0$ the GWT degenerates on the *ordinary Weyl transform* [Weyl (1950); Mizrahi (1975)].

We derive now some properties of the GWT. One should first notice that some specific mappings can be readily associated with the ordering of the noncommuting operators, Q and P, entering the structure of the composite operator $A(Q, P)$. For instance, for $\alpha = -1/2$ Eq.(1.52) becomes

$$\begin{aligned}
a_{\frac{1}{2}}(q,p) &= \int_{-\infty}^{+\infty} d^N\tau \, e^{\frac{i}{\hbar}\tau\cdot p} \, \langle q \big| A(Q,P) \big| q+\tau \rangle \\
&= \int_{-\infty}^{+\infty} d^N p' \int_{-\infty}^{+\infty} d^N\tau \, e^{\frac{i}{\hbar}\tau\cdot p} \, \langle q \big| A(Q,P) \big| p' \rangle \langle p' \big| q+\tau \rangle \\
&= (2\pi\hbar)^{-\frac{N}{2}} \int_{-\infty}^{+\infty} d^N p' \int_{-\infty}^{+\infty} d^N\tau \, e^{\frac{i}{\hbar}[\tau\cdot p - \tau\cdot p' - q\cdot p']} \langle q \big| A(Q,P) \big| p' \rangle \\
&= (2\pi\hbar)^{+\frac{N}{2}} \, e^{-\frac{i}{\hbar} q\cdot p} \, \langle q \big| A(Q,P) \big| p \rangle \\
&= (2\pi\hbar)^N \, \langle p \big| q \rangle \langle q \big| A(Q,P) \big| p \rangle \,.
\end{aligned} \tag{1.54}$$

Thus, up to a multiplicative factor, the function $a_{\frac{1}{2}}(q,p)$ can be read from the operator $A(Q, P)$ after all the Q's be reordered to the left of the P's by using the commutation rules in Eq.(1.27). This is the *standard ordering.* Similarly, $a_{-\frac{1}{2}}(q,p)$ is correlated with the form assumed by the operator $A(Q, P)$ after all the P's be reordered to the left of the Q's, i.e., the *anti-standard ordering.*

We focus, afterwards, on the double mapping

$$A \overset{\alpha}{\longrightarrow} a_\alpha \overset{\beta}{\longrightarrow} A_{\alpha,\beta} \,. \tag{1.55}$$

From Eqs.(1.50) and (1.51) we obtain

$$\begin{aligned}
A_{\alpha,\beta}(Q,P) &= \int_{-\infty}^{+\infty} d^N q \int_{-\infty}^{+\infty} d^N k \, a_\alpha(q,k) \, \Delta_\beta(Q-q, K-k) \\
&= (2\pi\hbar)^{-N} \int_{-\infty}^{+\infty} d^N q \int_{-\infty}^{+\infty} d^N k \int_{-\infty}^{+\infty} d^N u \int_{-\infty}^{+\infty} d^N v \, e^{\frac{i}{\hbar} k\cdot(v-u)} \bigg| q - \left(\frac{1}{2}+\beta\right) u \bigg\rangle \\
&\times \bigg\langle q - \left(\frac{1}{2}-\alpha\right) v \bigg| \, A(Q,P) \bigg| q + \left(\frac{1}{2}+\alpha\right) v \bigg\rangle \bigg\langle q + \left(\frac{1}{2}-\beta\right) u \bigg| \,.
\end{aligned} \tag{1.56}$$

The integration on k gives rise to $\delta^{(N)}(v-u)$ which makes the integration on v straightforward. Hence,

$$A_{\alpha,\beta}(Q,P) = \int_{-\infty}^{+\infty} d^N q \int_{-\infty}^{+\infty} d^N u \left| q - \left(\frac{1}{2}+\beta\right) u \right\rangle \\ \times \left\langle q - \left(\frac{1}{2}-\alpha\right) u \right| A(Q,P) \left| q + \left(\frac{1}{2}+\alpha\right) u \right\rangle \\ \times \left\langle q + \left(\frac{1}{2}-\beta\right) u \right| . \tag{1.57}$$

Since

$$\left\langle q - \left(\frac{1}{2}-\alpha\right) u \right| = \left\langle q - \left(\frac{1}{2}+\beta\right) u \right| e^{\frac{i}{\hbar}(\alpha+\beta)u\cdot P} , \tag{1.58a}$$

$$\left| q + \left(\frac{1}{2}+\alpha\right) u \right\rangle = e^{-\frac{i}{\hbar}(\alpha+\beta)u\cdot P} \left| q + \left(\frac{1}{2}-\beta\right) u \right\rangle , \tag{1.58b}$$

we can rewrite Eq.(1.57) as

$$A_{\alpha,\beta}(Q,P) = \int_{-\infty}^{+\infty} d^N q \int_{-\infty}^{+\infty} d^N u \left| q - \left(\frac{1}{2}+\beta\right) u \right\rangle \left\langle q - \left(\frac{1}{2}+\beta\right) u \right| \\ \times e^{\frac{i}{\hbar}(\alpha+\beta)u\cdot P} A(Q,P)\, e^{-\frac{i}{\hbar}(\alpha+\beta)u\cdot P} \\ \times \left| q + \left(\frac{1}{2}-\beta\right) u \right\rangle \left\langle q + \left(\frac{1}{2}-\beta\right) u \right| . \tag{1.59}$$

We now change integration variables as follows

$$q_i \longrightarrow q_i' = q_i - \left(\frac{1}{2}+\beta\right) u_i , \tag{1.60a}$$

$$u_i \longrightarrow q_i'' = q_i + \left(\frac{1}{2}-\beta\right) u_i . \tag{1.60b}$$

Concerning the integration measure we have that

$$d^N q d^N u = J\left(\frac{q\,,\,u}{q'\,,\,q''}\right) d^N q'\, d^N q'' , \tag{1.61}$$

where J is the Jacobian of the transformation. In order to compute J we first look at the $2N \times 2N$ square matrix $\|J\|$ whose elements,

$$J_{ik} = \begin{cases} \delta_{ik}\left(\frac{1}{2}-\beta\right), & 1 \le i \le N, \quad 1 \le k \le N \\ \delta_{ik}\left(\frac{1}{2}+\beta\right), & 1 \le i \le N, \quad N+1 \le k \le 2N \\ -\delta_{ik}, & N+1 \le i \le 2N, \quad 1 \le k \le N \\ +\delta_{ik}, & N+1 \le i \le 2N, \; N+1 \le k \le 2N \end{cases} , \tag{1.62}$$

can be gathered from Eq.(1.60). We recall that $\det \|J\|$ will remain unaltered if we substitute the first column ($k = 1$) of $\|J\|$ by the sum of columns $(k = 1) + (k = N + 1)$. The elements of the new first column are

$$J_{i1} = +\delta_{i1}, \quad 1 \leq i \leq 2N\,. \tag{1.63}$$

Then, $\det \|J\|$ equals the cofactor of the element $J_{11} = +1$. By pursuing this systematics further on we end up with

$$\det \|J\| = 1 \Longrightarrow J \equiv |\det \|J\|| = 1\,. \tag{1.64}$$

Therefore, Eq.(1.59) boils down to

$$\begin{aligned} A_{\alpha,\beta}(Q,P) &= \int_{-\infty}^{+\infty} d^N q' \int_{-\infty}^{+\infty} d^N q'' \\ &\times |q'\rangle \langle q'| \, e^{-\frac{i}{\hbar}(\alpha+\beta)(q'-q'')\cdot P} A(Q,P) e^{+\frac{i}{\hbar}(\alpha+\beta)(q'-q'')\cdot P} \\ &\times |q''\rangle \langle q''|\,, \end{aligned} \tag{1.65}$$

from which it follows that

$$A_{\alpha,\beta}(Q,P) = A_{\beta,\alpha}(Q,P)\,, \tag{1.66a}$$

$$A_{\alpha,\beta}(Q,P) = A_{0,\alpha+\beta}(Q,P) = A_{\alpha+\beta,0}(Q,P)\,, \tag{1.66b}$$

$$A_{\alpha,-\alpha}(Q,P) = A_{-\alpha,\alpha}(Q,P) = A_{0,0}(Q,P) = A(Q,P)\,, \tag{1.66c}$$

$$A_{\alpha,\beta}(Q,P) = A_{\alpha+\gamma,\beta-\gamma}(Q,P)\,. \tag{1.66d}$$

Another double mapping of interest is

$$a \xrightarrow{\alpha} A_\alpha \xrightarrow{\beta} a_{\alpha,\beta}\,. \tag{1.67}$$

With the help of Eq.(1.52) it can be written as

$$\begin{aligned} & a_{\alpha,\beta}(q,p) \\ &= \int_{-\infty}^{+\infty} d^N \tau \, e^{\frac{i}{\hbar}\tau\cdot p} \\ &\times \left\langle q - \left(\frac{1}{2} - \beta\right)\tau \right| A_\alpha(Q,P) \left| q + \left(\frac{1}{2} + \beta\right)\tau \right\rangle, \end{aligned} \tag{1.68}$$

which on account of Eqs.(1.50) and (1.51) becomes

$$a_{\alpha,\beta}(q,p) = (2\pi\hbar)^{-N}\int_{-\infty}^{+\infty} d^N v \int_{-\infty}^{+\infty} d^N k \int_{-\infty}^{+\infty} d^N u \int_{-\infty}^{+\infty} d^N r$$
$$\times e^{\frac{i}{\hbar}v\cdot p}\, e^{-\frac{i}{\hbar}r\cdot k}\left\langle q-\left(\frac{1}{2}-\beta\right)v\middle|u-\left(\frac{1}{2}+\alpha\right)r\right\rangle$$
$$\times\left\langle u+\left(\frac{1}{2}-\alpha\right)r\middle|q+\left(\frac{1}{2}+\beta\right)v\right\rangle a(u,k)\,. \tag{1.69}$$

The normalization condition in Eq.(1.30) and

$$\delta^{(N)}[a(q-q')] = \left(\frac{1}{|a|}\right)^N \delta^{(N)}(q'-q) \tag{1.70}$$

imply that

$$\left\langle q-\left(\frac{1}{2}-\beta\right)v\middle|u-\left(\frac{1}{2}+\alpha\right)r\right\rangle$$
$$= \frac{1}{\left|\frac{1}{2}-\beta\right|^N}\,\delta^{(N)}\left(v-\frac{q-u+\frac{1}{2}r+\alpha r}{\frac{1}{2}-\beta}\right), \tag{1.71a}$$
$$\left\langle u+\left(\frac{1}{2}-\alpha\right)r\middle|q+\left(\frac{1}{2}+\beta\right)v\right\rangle$$
$$= \frac{1}{\left|\frac{1}{2}+\beta\right|^N}\,\delta^{(N)}\left(v-\frac{u-q+\frac{1}{2}r-\alpha r}{\frac{1}{2}+\beta}\right). \tag{1.71b}$$

By returning with Eqs.(1.71) into (1.69) and after carrying out the integral on v we find

$$a_{\alpha,\beta}(q,p) = (2\pi\hbar)^{-N}\int_{-\infty}^{+\infty} d^N k \int_{-\infty}^{+\infty} d^N u \int_{-\infty}^{+\infty} d^N r$$
$$\times \exp\left[\frac{i}{\hbar}\left(\frac{q-u+\frac{1}{2}r+\alpha r}{\frac{1}{2}-\beta}\right)\right]\exp\left(-\frac{i}{\hbar}r\cdot k\right)$$
$$\times\,\delta^{(N)}\left[q-u+(\alpha+\beta)\,r\right]\,a(u,k)\,. \tag{1.72}$$

The u integral is straightforward and yields

$$a_{\alpha,\beta}(q,p) = (2\pi\hbar)^{-N}$$
$$\times\int_{-\infty}^{+\infty} d^N k \int_{-\infty}^{+\infty} d^N r\, e^{\frac{i}{\hbar}(p-k)\cdot r}\, a\left[q+(\alpha+\beta)r,k\right], \tag{1.73}$$

leading to

$$a_{\alpha,\beta}(q,p) = a_{\beta,\alpha}(q,p)\,, \tag{1.74a}$$
$$a_{\alpha,\beta}(q,p) = a_{0,\alpha+\beta}(q,p) = a_{\alpha+\beta,0}(q,p)\,, \tag{1.74b}$$
$$a_{\alpha,-\alpha}(q,p) = a_{-\alpha,\alpha}(q,p) = a_{0,0}(q,p) = a(q,q)\,, \tag{1.74c}$$
$$a_{\alpha,\beta}(q,p) = a_{\alpha+\gamma,\beta-\gamma}(q,p)\,, \tag{1.74d}$$

which are the counterparts of Eqs.(1.66).

More complex mappings are discussed further below. We now look at

$$a \xrightarrow{\alpha_1} A_{\alpha_1} \xrightarrow{\alpha_2} a_{\alpha_1,\alpha_2} \cdots \xrightarrow{\alpha_n} A_{\alpha_1,\alpha_2,\ldots,\alpha_n} \tag{1.75}$$

where $|\alpha_i| \leq 1/2, i = 1, 2, \ldots, n$ and n is an *odd* integer. From Eqs.(1.66b) and (1.66c) we find

$$A_{\alpha_1,\alpha_2,\ldots,\alpha_{n-2},\alpha_{n-1},\alpha_n} = A_{\alpha_1,\alpha_2,\ldots,\alpha_{n-2},\alpha_{n-1}+\alpha_n,0}$$
$$= A_{\alpha_1,\alpha_2,\ldots,\alpha_{n-2}+\alpha_{n-1}+\alpha_n,0,0} = A_{\alpha_1,\alpha_2,\ldots,\alpha_{n-2}+\alpha_{n-1}+\alpha_n}\,, \tag{1.76}$$

which ultimately ends at

$$A_{\alpha_1,\alpha_2,\ldots,\alpha_{n-2},\alpha_{n-1},\alpha_n} = A_{\alpha_1+\alpha_2+\ldots+\alpha_{n-2}+\alpha_{n-1}+\alpha_n}\,. \tag{1.77}$$

Similarly, Eqs.(1.74b) and (1.74c) lead to

$$a_{\alpha_1,\alpha_2,\ldots,\alpha_{n-2},\alpha_{n-1},\alpha_n} = a_{\alpha_1+\alpha_2+\ldots+\alpha_{n-2}+\alpha_{n-1}+\alpha_n}\,. \tag{1.78}$$

For *even* n the following holds true

$$A_{\alpha_1,\alpha_2,\ldots,\alpha_{n-2},\alpha_{n-1},\alpha_n} = A_{0,\alpha_1+\alpha_2+\ldots+\alpha_{n-2}+\alpha_{n-1}+\alpha_n}\,, \tag{1.79}$$

$$a_{\alpha_1,\alpha_2,\ldots,\alpha_{n-2},\alpha_{n-1},\alpha_n} = a_{0,\alpha_1+\alpha_2+\ldots+\alpha_{n-2}+\alpha_{n-1}+\alpha_n}\,. \tag{1.80}$$

Another relevant relationship is the link between the GWT and the ordinary Weyl transform ($\alpha = 0$). To find it, we shall begin from Eq.(1.73) written as

$$a_{0,\alpha,\beta}(q,p) = (2\pi\hbar)^{-N}$$
$$\times \int_{-\infty}^{+\infty} d^N k \int_{-\infty}^{+\infty} d^N r\, e^{\frac{i}{\hbar}(p-k)\cdot r}\, a_0\left[q + (\alpha+\beta)r, k\right]\,, \tag{1.81}$$

or, equivalently,

$$a_{\alpha+\beta}(q,p) = (2\pi\hbar)^{-N} \times \int_{-\infty}^{+\infty} d^N k \int_{-\infty}^{+\infty} d^N r \, e^{\frac{i}{\hbar}(p-k)\cdot r} \, a_0 \left[q + (\alpha+\beta) r, k\right] , \tag{1.82}$$

since in view of Eq.(1.78)

$$a_{0,\alpha,\beta}(q,p) = a_{0+\alpha+\beta} = a_{\alpha+\beta}(q,p) \, . \tag{1.83}$$

Moreover, for $\beta = 0$ Eq.(1.82) reduces to

$$a_\alpha(q,p) = (2\pi\hbar)^{-N} \int_{-\infty}^{+\infty} d^N k \int_{-\infty}^{+\infty} d^N r \, e^{\frac{i}{\hbar}(p-k)\cdot r} \, a_0 \left(q + \alpha r, k\right) \, . \tag{1.84}$$

We shall assume that $a_0(q,p)$ is analytic in a neighborhood of the origin. By invoking its McLauring expansion we can then write

$$\begin{aligned} a_0\left(q+\alpha r, k\right) &= \sum_{n=0}^{\infty} \frac{\alpha^n}{n!} r_{i_1} \ldots r_{i_n} \frac{\partial^n a_0(q,k)}{\partial q_{i_1} \ldots \partial q_{i_n}} \\ &= e^{\alpha r \cdot \frac{\partial}{\partial q}} \, a_0(q,k) \, , \end{aligned} \tag{1.85}$$

which makes possible to cast Eq.(1.84) as

$$\begin{aligned} a_\alpha(q,p) &= (2\pi\hbar)^{-N} \int_{-\infty}^{+\infty} d^N k \int_{-\infty}^{+\infty} d^N r \, e^{\frac{i}{\hbar}(p-k)\cdot r} \, e^{\alpha r \cdot \frac{\partial}{\partial q}} \, a_0(q,k) \\ &= \int_{-\infty}^{+\infty} d^N r \, e^{\frac{i}{\hbar} r \cdot \left(p - i\hbar\alpha \frac{\partial}{\partial q}\right)} \, g_0(q,r) \, , \end{aligned} \tag{1.86}$$

where

$$g_0(q,r) = (2\pi\hbar)^{-N} \int_{-\infty}^{+\infty} d^N k \, e^{-\frac{i}{\hbar} k \cdot r} \, a_0(q,k) \tag{1.87}$$

is the Fourier transform of $a_0(q,k)$. Accordingly, the right-hand side of Eq.(1.86) is the inverse Fourier transform of $g_0(q,r)$,

$$a_0(q,k) = \int_{-\infty}^{+\infty} d^N r \, e^{\frac{i}{\hbar} k \cdot r} \, g_0(q,r) \, , \tag{1.88}$$

evaluated at $k = p - i\hbar\alpha\frac{\partial}{\partial q}$, i.e.,

$$a_\alpha(q,p) = a_0\left(q, p - i\hbar\alpha\frac{\partial}{\partial q}\right) = e^{-i\hbar\alpha\frac{\partial}{\partial q}\cdot\frac{\partial}{\partial p}}a_0(q,p)\,, \tag{1.89}$$

which is the desired expression. Thus, the ordinary Weyl transform can be used as input for computing the GWT.

The last part of this section deals with the restrictions imposed on the structure of $a_\alpha(q,p)$ by the requirement $A^\dagger(Q,P) = A(Q,P)$ as well as those imposed on the structure of $A_\alpha(Q,P)$ by the requirement $a^*(q,p) = a(q,p)$.

On the one hand, from Eq.(1.52) and after changing the integration variable $\tau \to -\tau$ we find

$$\begin{aligned} &a^\star_\alpha(q,p) \\ &\equiv \int_{-\infty}^{+\infty} d^N\tau\, e^{\frac{i}{\hbar}\tau\cdot p} \\ &\times \left\langle q - \left(\frac{1}{2}+\alpha\right)\tau \right| A^\dagger(Q,P) \left| q + \left(\frac{1}{2}-\alpha\right)\tau \right\rangle. \end{aligned} \tag{1.90}$$

Hence,

$$A^\dagger(Q,P) = A(Q,P) \Longrightarrow a^\star_\alpha(q,p) = a_{-\alpha}(q,p)\,. \tag{1.91}$$

Then, only for $\alpha = 0$ (ordinary Weyl transform) Hermitian operators map one-one onto real functions.

On the other hand, by taking the Hermitian adjoint on both sides of Eq.(1.50) we are led to

$$A^\dagger_\alpha(Q,P) \equiv \int_{-\infty}^{+\infty} d^N p \int_{-\infty}^{+\infty} d^N q\, a^\star(q,p)\,\Delta^\dagger_\alpha(Q-q,P-p)\,. \tag{1.92}$$

In addition, from Eq.(1.51) it follows that

$$\begin{aligned} &\Delta^\dagger_\alpha(Q-q,P-p) \equiv (2\pi\hbar)^{-N} \\ &\int_{-\infty}^{+\infty} d^N\tau\, e^{+\frac{i}{\hbar}\tau\cdot p} \left| q + \left(\frac{1}{2}-\alpha\right)\tau \right\rangle \left\langle q - \left(\frac{1}{2}+\alpha\right)\tau \right| \\ &= \Delta_{-\alpha}(Q-q,P-p)\,. \end{aligned} \tag{1.93}$$

Thus, by substituting Eq.(1.93) into Eq.(1.92) we get

$$A^\dagger_\alpha(Q,P) \equiv \int_{-\infty}^{+\infty} d^N p \int_{-\infty}^{+\infty} d^N q\, a^\star(q,p)\,\Delta_{-\alpha}(Q-q,P-p)\,. \tag{1.94}$$

Therefore,

$$a^\star(q,p) = a(q,p) \Longrightarrow A_\alpha^\dagger(Q,P) = A_{-\alpha}(Q,P)\,. \tag{1.95}$$

The distinctive role played by the ordinary Weyl transform should again be noticed. Indeed, only for $\alpha = 0$ real functions map one-one onto Hermitian operators.

1.2.2 *The ordering problem*

We now look for gaining insight about the origin of the α dependence. We shall do this through examples.

We first analyze $a_\alpha(q,p)$ when the input is the operator

$$A(Q,P) = c\,F(Q) + d\,g(P)\,, \tag{1.96}$$

where c and d are complex constants. Notice that the operator in Eq.(1.96) does not contain products of noncommuting factors. In order to find the corresponding GWT we shall begin from Eq.(1.52)

$$\begin{aligned} &a_\alpha(q,p) \\ &= \int_{-\infty}^{+\infty} d^N\tau\, e^{\frac{i}{\hbar}\tau\cdot p}\, c\left\langle q-\left(\frac{1}{2}-\alpha\right)\tau\right| f(Q)\left|q+\left(\frac{1}{2}+\alpha\right)\tau\right\rangle \\ &+ \int_{-\infty}^{+\infty} d^N\tau e^{\frac{i}{\hbar}\tau\cdot p} d\left\langle q-\left(\frac{1}{2}-\alpha\right)\tau\right| g(P)\left|q+\left(\frac{1}{2}+\alpha\right)\tau\right\rangle, \end{aligned} \tag{1.97}$$

which asserts that the operation of taking the GWT is linear. Furthermore, we can verify that

$$\begin{aligned} &\left\langle q-\left(\frac{1}{2}-\alpha\right)\tau\right| f(Q)\left|q+\left(\frac{1}{2}+\alpha\right)\tau\right\rangle \\ &= f\left[q+\left(\frac{1}{2}+\alpha\right)\tau\right]\,\delta^{(N)}(\tau)\,, \end{aligned} \tag{1.98}$$

whereas

$$\begin{aligned} &\left\langle q-\left(\frac{1}{2}-\alpha\right)\tau\right| g(P)\left|q+\left(\frac{1}{2}+\alpha\right)\tau\right\rangle \\ &= \int_{-\infty}^{+\infty} d^Np'\, g\,(p')\left\langle q-\left(\frac{1}{2}-\alpha\right)\tau\right|p'\right\rangle\left\langle p'\right|q+\left(\frac{1}{2}+\alpha\right)\tau\right\rangle \\ &= \frac{1}{(2\pi\hbar)^N}\int_{-\infty}^{+\infty} d^Np'\, g(p')\, e^{-\frac{i}{\hbar}p'\cdot\tau}\,. \end{aligned} \tag{1.99}$$

By replacing Eqs.(1.98) and (1.99) into (1.97) we find

$$a_\alpha(q,p) = c\,f(q) + d\,g(p)\,. \tag{1.100}$$

Therefore, when $A(Q,P)$ does not contain products of noncommuting operators, the corresponding GWT *does not depend* on α.

Let us now ask about the dependence on α of the operators $A_\alpha(Q,P)$ originating from the phase space function

$$a(q,p) = q\,p^2\,. \tag{1.101}$$

Now we begin from Eqs.(1.50) and (1.51). They yield

$$\begin{aligned} A_\alpha(Q,P) = (2\pi\hbar)^{-1} \int_{-\infty}^{+\infty} dq \int_{-\infty}^{+\infty} dp \int_{-\infty}^{+\infty} d\tau\, q\,p^2 \\ \times\, e^{-\frac{i}{\hbar}\tau p} \left| q - \left(\frac{1}{2}+\alpha\right)\tau \right\rangle \left\langle q + \left(\frac{1}{2}-\alpha\right)\tau \right| . \end{aligned} \tag{1.102}$$

However,

$$\begin{aligned} &\left| q - \left(\frac{1}{2}+\alpha\right)\tau \right\rangle q \left\langle q + \left(\frac{1}{2}-\alpha\right)\tau \right| \\ &= e^{+\frac{i}{\hbar}\left(\frac{1}{2}+\alpha\right)\tau P} |q\rangle\, q\, \langle q| e^{+\frac{i}{\hbar}\left(\frac{1}{2}-\alpha\right)\tau P} \end{aligned} \tag{1.103}$$

which leads to

$$\begin{aligned} &\int_{-\infty}^{+\infty} dq \left| q - \left(\frac{1}{2}+\alpha\right)\tau \right\rangle q \left\langle q + \left(\frac{1}{2}-\alpha\right)\tau \right| \\ &= e^{+\frac{i}{\hbar}\left(\frac{1}{2}+\alpha\right)\tau P} \left(\int_{-\infty}^{+\infty} dq |q\rangle\, q\, \langle q| \right) e^{+\frac{i}{\hbar}\left(\frac{1}{2}-\alpha\right)\tau P} \\ &= e^{+\frac{i}{\hbar}\left(\frac{1}{2}+\alpha\right)\tau P}\, Q\, e^{+\frac{i}{\hbar}\left(\frac{1}{2}-\alpha\right)\tau P} \\ &= e^{+\frac{i}{\hbar}\left(\frac{1}{2}+\alpha\right)\tau P} \underbrace{e^{+\frac{i}{\hbar}\left(\frac{1}{2}-\alpha\right)\tau P}\, e^{-\frac{i}{\hbar}\left(\frac{1}{2}-\alpha\right)\tau P}}_{I}\, Q\, e^{+\frac{i}{\hbar}\left(\frac{1}{2}-\alpha\right)\tau P} \\ &= e^{+\frac{i}{\hbar}\tau P} \left[Q - \left(\frac{1}{2}-\alpha\right)\tau I \right] . \end{aligned} \tag{1.104}$$

For arriving at this result we took into account the spectral resolution of the position operator,

$$Q = \int_{-\infty}^{+\infty} dq |q\rangle\, q\, \langle q|\,, \tag{1.105}$$

as well as the translation property

$$e^{-\frac{i}{\hbar}\left(\frac{1}{2}-\alpha\right)\tau P}\, Q\, e^{+\frac{i}{\hbar}\left(\frac{1}{2}-\alpha\right)\tau P} = Q - \left(\frac{1}{2}-\alpha\right) \tau I \,. \tag{1.106}$$

From Eqs.(1.104) and Eq.(1.102) we conclude that

$$\begin{aligned} A_\alpha(Q,P) = (2\pi\hbar)^{-1} \int_{-\infty}^{+\infty} dp \int_{-\infty}^{+\infty} d\tau\, p^2\, e^{-\frac{i}{\hbar}\tau p}\, e^{\frac{i}{\hbar}\tau P} \\ \times \left[Q - \left(\frac{1}{2}-\alpha\right) \tau I \right] . \end{aligned} \tag{1.107}$$

After rearrangements, this last expression can be cast

$$\begin{aligned} A_\alpha(Q,P) = P^2\, Q - (2\pi\hbar)^{-1} \left(\frac{1}{2}-\alpha\right) \int_{-\infty}^{+\infty} dp \int_{-\infty}^{+\infty} d\tau \int_{-\infty}^{+\infty} dp' \\ \times\ p^2\, \tau\, e^{\frac{i}{\hbar}\tau(p-p')}\, |p'\rangle\langle p'| \,, \end{aligned} \tag{1.108}$$

which, with the help of the identity

$$(2\pi\hbar)^{-1} \int_{-\infty}^{+\infty} d\tau\, \tau\, e^{\frac{i}{\hbar}\tau(p-p')} \equiv -i\hbar \frac{\partial \delta\,(p-p')}{\partial p'} \,, \tag{1.109}$$

can be shown to be equivalent to

$$A_\alpha(Q,P) = P^2\, Q + i\hbar\, (1-2\,\alpha)\, P \,. \tag{1.110}$$

Let us now analyze special cases. For instance,

$$\alpha = +\frac{1}{2} \Longrightarrow A_{+\frac{1}{2}}(Q,P) = P^2 Q \,. \tag{1.111}$$

Thus, the operator $A_{+\frac{1}{2}}(Q,P)$ can be read from the classical function in Eq.(1.101) according to the following rules: i) to substitute $q \longrightarrow Q$ and $p \longrightarrow P$, ii) to locate all operators P on the left of the Q's. This last part of the prescription has already been referred to as the *anti-standard ordering*. On the other hand, for $\alpha = -\frac{1}{2}$ rule i) goes as above while rule ii) requires switching from anti-standard to *standard ordering*, namely,

$$\alpha = -\frac{1}{2} \Longrightarrow A_{-\frac{1}{2}}(Q,P) = P^2 Q + 2i\hbar\, P = Q P^2 \,. \tag{1.112}$$

Another case deserving attention is

$$\alpha = 0 \Longrightarrow A_0 = P^2 Q + i\hbar P = \frac{1}{3}\left(P^2 Q + PQP + QP^2\right) . \tag{1.113}$$

Notice that the operators Q and P on the right-hand side of this last equation have been ordered in accordance with the *Weyl symmetrization rule.*

Thus, each value of α becomes associated with an *ordering prescription.*

1.3 The phase space path integral

We recourse to the GWT for obtaining a phase space path integral representation for the propagator.

We begin by *slicing* the time interval $[t_f, t_i]$ into $(m+1)$ subintervals of equal size, i.e.,

$$t_i = t_0 < t_1 < t_2 < \ldots < t_{m-1} < t_m < t_{m+1} = t_f . \tag{1.114}$$

Next, we introduce the *short time propagator* (STP)

$$\begin{aligned} & K(q_{j+1}, t_{j+1}; q_j, t_j) \\ & \equiv {}_H\langle q_{j+1}, t_{j+1} | q_j, t_j \rangle_H \\ & = \langle q_{j+1} | \, e^{-\frac{i}{\hbar} H(Q,P)(t_{j+1}-t_j)} \, | q_j \rangle . \end{aligned} \tag{1.115}$$

One can notice that the propagator in Eq.(1.44) can be written as a product of STP's. In fact, when the expression giving the spectral resolution of the identity operator in terms of Heisenberg picture position eigenvectors,

$$I = \int_{-\infty}^{+\infty} d^N q \, |q, t\rangle_H \, {}_H\langle q, t| , \tag{1.116}$$

is inserted m times in Eq.(1.44) we find

$$\begin{aligned} K\left(q_f, t_f; q_i, t_i\right) &= \int_{-\infty}^{+\infty} d^N q_1 \ldots d^N q_m \, {}_H\langle q_f, t_f | q_m, t_m \rangle_H \, {}_H\langle q_m, t_m | \\ & \times \cdots | q_1, t_1 \rangle_H \, {}_H\langle q_1, t_1 | q_i, t_i \rangle_H \\ &= \int_{-\infty}^{+\infty} d^N q_1 \ldots d^N q_m \prod_{j=0}^{m} K(q_{j+1}, t_{j+1}; q_j, t_j) . \end{aligned} \tag{1.117}$$

The main object is, then, the STP on which we shall concentrate now. To simplify the writing we introduce the definition

$$\epsilon \equiv t_{j+1} - t_j \,. \tag{1.118}$$

The convergence of the series expansion of the STP in powers of ϵ is secured by choosing a sufficiently small time slice size. The limit $\epsilon \to 0$ will always be taken at the end of the calculations. We, then, approximate the STP by retaining the terms up to the first order in ϵ only, i.e.,

$$\begin{aligned} K(q_{j+1}, t_{j+1}; q_j, t_j) &= \langle q_{j+1} | \, e^{-\frac{i}{\hbar} H(Q,P)(t_{j+1}-t_j)} \, | q_j \rangle \\ \simeq \langle q_{j+1} | & \left(I - \frac{i}{\hbar} \, \epsilon \, H \right) | q_j \rangle \,. \end{aligned} \tag{1.119}$$

The GWT enters the game by invoking Eq.(1.66c), i.e.,

$$\left[I - \frac{i}{\hbar} \, \epsilon \, H(Q,P) \right] = \left[I - \frac{i}{\hbar} \, \epsilon \, H(Q,P) \right]_{\alpha,-\alpha} \,, \tag{1.120}$$

which along with the definitions in Eqs.(1.50) and (1.52) leads to

$$\begin{aligned} \left[I - \frac{i}{\hbar} \, \epsilon \, H(Q,P) \right]_{\alpha,-\alpha} &= \int_{-\infty}^{+\infty} d^N p \int_{-\infty}^{+\infty} d^N q \left[1 - \frac{i}{\hbar} \, \epsilon \, h_\alpha(q,p) \right] \\ \times \Delta_{-\alpha} \left(Q - q, P - p \right) \,, & \end{aligned} \tag{1.121}$$

where

$$\begin{aligned} & h_\alpha(q,p) \\ & = \int_{-\infty}^{+\infty} d^N \tau e^{\frac{i}{\hbar} \tau \cdot p} \\ & \times \left\langle q - \left(\frac{1}{2} - \alpha \right) \tau \right| H(Q,P) \left| q + \left(\frac{1}{2} + \alpha \right) \tau \right\rangle \end{aligned} \tag{1.122}$$

is the GWT of the Hamiltonian operator. By returning with Eq.(1.121) into (1.119) we obtain

$$\begin{aligned} K(q_{j+1}, t_{j+1}; q_j, t_j) &\simeq \int_{-\infty}^{+\infty} d^N p \int_{-\infty}^{+\infty} d^N q \left[1 - \frac{i}{\hbar} \, \epsilon \, h_\alpha(q,p) \right] \\ \times \langle q_{j+1} | \Delta_{-\alpha} \left(Q - q, P - p \right) | q_j \rangle \,. & \end{aligned} \tag{1.123}$$

In order to compute the matrix element on the right hand side of this last equation we take advantage of Eq.(1.51). We then find

$$\begin{aligned}
\langle q_{j+1}|\Delta_{-\alpha}\left(Q-q,P-p\right)|q_j\rangle &= (2\pi\hbar)^{-N}\int_{-\infty}^{+\infty} d^N\tau\, e^{-\frac{i}{\hbar}\tau\cdot p} \\
&\times \left\langle q_{j+1}\middle| q-\left(\frac{1}{2}-\alpha\right)\tau\right\rangle \left\langle q+\left(\frac{1}{2}+\alpha\right)\tau \middle| q_j\right\rangle \\
&= (2\pi\hbar)^{-N}\int_{-\infty}^{+\infty} d^N\tau\, e^{-\frac{i}{\hbar}\tau\cdot p}\,\delta^{(N)}\left[q_{j+1}-q+\left(\frac{1}{2}-\alpha\right)\tau\right] \\
&\times \delta^{(N)}\left[q_j-q-\left(\frac{1}{2}+\alpha\right)\tau\right] \\
&= (2\pi\hbar)^{-N}\, e^{\frac{i}{\hbar}p\cdot(q_{j+1}-q_j)}\,\delta^{(N)}[q-q_j(\alpha)]\,,
\end{aligned} \tag{1.124}$$

where

$$q_j(\alpha) \equiv \left(\frac{1}{2}-\alpha\right)q_j+\left(\frac{1}{2}+\alpha\right)q_{j+1}\,. \tag{1.125}$$

By going back with Eq.(1.124) into Eq.(1.123) and after performing the q-integration we arrive at

$$\begin{aligned}
K(q_{j+1},t_{j+1};q_j,t_j) &\simeq (2\pi\hbar)^{-N}\int_{-\infty}^{+\infty} d^N p \\
&\times \exp\left\{\frac{i}{\hbar}\left[p\,\frac{(q_{j+1}-q_j)}{(t_{j+1}-t_j)}-h_\alpha(q_j(\alpha),p)\right](t_{j+1}-t_j)\right\}\,.
\end{aligned} \tag{1.126}$$

Notice that the factor $1-\frac{i}{\hbar}\,\epsilon\, h_\alpha(q,p)$, which shows up on the right hand side of Eq.(1.123), was replaced by $\exp[-\frac{i}{\hbar}\epsilon h_\alpha(q,p)]$. Qualitatively, this is the *inverse* of the operation at the beginning of the STP computation where the *operator* $\exp[-\frac{i}{\hbar}\epsilon H(Q,P)]$ was approximated by $1-\frac{i}{\hbar}\,\epsilon\, H(Q,P)$. The computation of the STP is now complete.

By plugging back Eq.(1.126) into Eq.(1.117) and after carrying out the limiting operations

$$\epsilon \longrightarrow 0\,, \qquad m \longrightarrow \infty\,, \qquad \sum_{j=0}^{m}(t_{j+1}-t_j) = t_f - t_i\,, \tag{1.127}$$

we arrive at the final form

$$K\left(q_f, t_f; q_i, t_i\right) = \int_{-\infty}^{+\infty} d^N q_1 \ldots d^N q_m \prod_{j=0}^{m} K(q_{j+1}, t_{j+1}; q_j, t_j)$$

$$= \lim_{m\to\infty} (2\pi\hbar)^{-N(m+1)} \int_{-\infty}^{+\infty} \left(\prod_{j=1}^{m} d^N q_j\right) \int_{-\infty}^{+\infty} \left(\prod_{j=0}^{m} d^N p_j\right)$$

$$\times\, e^{\frac{i}{\hbar}\sum_{j=0}^{m}\left\{\left[p_j\cdot\frac{(q_{j+1}-q_j)}{(t_{j+1}-t_j)} - h_\alpha(q_j(\alpha),p)\right](t_{j+1}-t_j)\right\}} . \qquad (1.128)$$

We point out that there are $m \times N$ q-integrals and $(m+1)\times N$ p-integrals. The problem of establishing the existence of the limit $m \to \infty$ is far from trivial. When this is the case, the integrand in Eq.(1.128) tends to a well-defined functional. The expression in Eq.(1.128), referred to as the *phase space path integral*, yields a quantum amplitude, the propagator, in terms of classical quantities. It can be succinctly written as

$$K\left(q_f, t_f; q_i, t_i\right)$$
$$= \int_\alpha \frac{[\mathcal{D}q][\mathcal{D}p]}{(2\pi\hbar)^N}\, e^{\frac{i}{\hbar}\int_{t_i}^{t_f} dt\left[p(t)\cdot\frac{dq(t)}{dt} - h_\alpha(q(t),p(t))\right]} , \qquad (1.129)$$

where

$$[\mathcal{D}q] \equiv \prod_{j=1}^{m} d^N q_j \qquad (1.130)$$

and

$$[\mathcal{D}p] \equiv \prod_{j=0}^{m} d^N p_j \qquad (1.131)$$

define the functional integration measures[5].

Some remarks are in order:

1) The dependence on α on the right hand side of Eq.(1.128) signalizes a potential lack of uniqueness. It is an intrinsic ambiguity pervading the definition of the phase space path integral via the time slicing procedure. This dependence originates from two sources, since the function $h_\alpha(q,p)$ is

[5]The derivation of the phase space path integral is due to Garrod [Garrod (1966)] . The introduction of path integrals in quantum mechanics is due to Dirac [Dirac (1933)] and it was later on developed by Feynman [Feynman (1948)].

to be evaluated at $q_j = q_j(\alpha)$. The subscript α, under the integral sign in Eq.(1.129), is to remind us about this fact. We are not aware about the existence of a proof securing the cancellation of α-dependent terms. Each case must be individually investigated.

2) The exponent in Eq.(1.128),

$$\frac{i}{\hbar}\left[p_j \cdot \frac{(q_{j+1} - q_j)}{(t_{j+1} - t_j)} - h_\alpha(q_j(\alpha), p)\right] (t_{j+1} - t_j) \,, \tag{1.132}$$

is not a pure imaginary quantity since, on general grounds (see Eq.(1.91)), the GWT does not map the Hermitian operator $H(Q, P)$ onto a real function. What we know is that $h_\alpha(q,p) = h^\star_\alpha(q,p)$ either in the case of $\alpha = 0$ or when $H(Q, P)$ does not contain products of noncommuting operators, among other possibilities. When $h_\alpha(q,p)$ is a real quantity, the contributions to the functional integral arising from different configurations differ only by a phase factor.

1.4 One dimensional harmonic oscillator

The harmonic oscillator is a physical system whose propagator $K(q_f, t_f; q_i, t_i)$ can be exactly computed. It allows for a throughout test of the systematics put forward in the previous section. To simplify the calculations we shall restrict ourselves to consider the one dimensional harmonic oscillator ($N = 1$).

1.4.1 *Computation of the propagator*

The Hamiltonian operator is

$$H(Q, P) = \frac{P^2}{2M} + \frac{1}{2} M \omega^2 Q^2 \,, \tag{1.133}$$

where M is the particle mass and ω the frequency of oscillation. Since the Hamiltonian is not afflicted by ordering ambiguities its GWT does not depend on α and reads (recall Eq.(1.100))

$$h_\alpha(q, p) = \frac{p^2}{2M} + \frac{1}{2} M \omega^2 q^2 \,. \tag{1.134}$$

By going back with Eq.(1.134) into (1.128) we get

$$K\left(q_f,t_f;q_i,t_i\right) = \lim_{m\to\infty} (2\pi\hbar)^{-(m+1)} \int_{-\infty}^{+\infty} \left(\prod_{j=1}^{m} dq_j\right) \int_{-\infty}^{+\infty} \left(\prod_{j=0}^{m} dp_j\right)$$
$$\times \exp\left\{\frac{i}{\hbar}\sum_{j=0}^{m} \epsilon\left[p_j \frac{(q_{j+1}-q_j)}{\epsilon} - \frac{p_j^2}{2M} - \frac{1}{2}\,M\,\omega^2\,q_j^2(\alpha)\right]\right\}. \tag{1.135}$$

The momentum integrals are not entangled; they can be performed at once with the help of Eq.(A.10). Indeed,

$$\int_{-\infty}^{+\infty} dp_j\, e^{-\frac{i}{\hbar}\,\epsilon\,\frac{p_j^2}{2M} + \frac{i}{\hbar}p_j(q_{j+1}-q_j)} = \sqrt{\frac{2\pi\hbar M}{i\epsilon}}\; e^{\frac{iM}{\hbar}\,\frac{(q_{j+1}-q_j)^2}{2\epsilon}}. \tag{1.136}$$

When replaced into Eq.(1.135) they give rise to

$$K\left(q_f,t_f;q_i,t_i\right) = \lim_{m\to\infty} \left(\frac{M}{2\pi\hbar i\epsilon}\right)^{\frac{(m+1)}{2}}$$
$$\times \int_{-\infty}^{+\infty} \left(\prod_{j=1}^{m} dq_j\right) e^{\frac{i}{\hbar}\sum_{j=0}^{m}\epsilon\left[\frac{1}{2}M\left(\frac{q_{j+1}-q_j}{\epsilon}\right)^2 - \frac{1}{2}\,M\,\omega^2\,q_j^2(\alpha)\right]}. \tag{1.137}$$

To calculate the q-integrals we start by taking into account the definition in Eq.(1.125). Afterwards, the exponent in Eq.(1.137) is rearranged as follows

$$\sum_{j=0}^{m}\left[\frac{M}{2\epsilon}\left(q_{j+1}-q_j\right)^2 - \frac{1}{2}\,\epsilon\,M\,\omega^2\,q_j^2(\alpha)\right]$$
$$= R(\alpha)\,q_f^2 + H(\alpha)\,q_i^2 + \sum_{j=1}^{m} t_j\,q_j + \sum_{j,k=1}^{m} q_j\,a_{jk}\,q_k\,. \tag{1.138}$$

Here,

$$R(\alpha) \equiv \frac{M}{2\epsilon}\left[1 - \epsilon^2\,\omega^2\left(\frac{1}{2}+\alpha\right)^2\right], \tag{1.139a}$$

$$H(\alpha) \equiv \frac{M}{2\epsilon}\left[1 - \epsilon^2\,\omega^2\left(\frac{1}{2}-\alpha\right)^2\right], \tag{1.139b}$$

$$G(\alpha) \equiv \frac{M}{2\epsilon}\left[-1 - \epsilon^2\,\omega^2\left(\frac{1}{4}-\alpha^2\right)\right], \tag{1.139c}$$

while

$$t_1 \equiv 2\,G(\alpha)\,q_i\,, \quad t_2 = \ldots = t_{m-1} \equiv 0\,, \quad t_m \equiv 2\,G(\alpha)\,q_f\,, \tag{1.140}$$

and

$$a_{jk} \equiv [R(\alpha) + H(\alpha)]\,\delta_{jk} + G(\alpha)\,(\delta_{j-1,k} + \delta_{j,k-1}) = a_{kj}\,. \tag{1.141}$$

We go back with Eq.(1.138) into Eq.(1.137). The q-integrals are now of the same class appearing in Eq.(A.48) and can be readily computed [6]. We end up with

$$\begin{aligned} K\left(q_f, t_f; q_i, t_i\right) &= \lim_{m\to\infty} \left(\frac{M}{2\pi\hbar i\epsilon}\right)^{\frac{(m+1)}{2}} e^{\frac{i}{\hbar}\left[R(\alpha)\,q_f^2 + H(\alpha)\,q_i^2\right]} \\ &\times \int_{-\infty}^{+\infty} \left(\prod_{j=1}^{m} dq_j\right) e^{\frac{i}{\hbar}\sum_{j=1}^{m} t_j q_j}\, e^{\frac{i}{\hbar}\sum_{j,k=1}^{m} q_j\, a_{jk}\, q_k} \\ &= \lim_{m\to\infty} \left(\frac{M}{2\pi\hbar i\epsilon}\right)^{\frac{(m+1)}{2}} e^{\frac{i}{\hbar}\left[R(\alpha)\,q_f^2 + H(\alpha)\,q_i^2 - \frac{c^2}{4}\right]} \\ &\times (i\pi\hbar)^{\frac{m}{2}}\,(\det A)^{-\frac{1}{2}}\,, \end{aligned} \tag{1.142}$$

where A symbolizes the operator whose matrix elements are $\{a_{jk}\}$. Moreover,

$$c^2 = \sum_{j,k=1}^{m} t_j\, g_{jk}\, t_k\,, \tag{1.143}$$

where $\{g_{jk}\}$ denotes the matrix elements of the operator A^{-1} (see Eq.(A.25)). Below, we introduce the operator D through its matrix elements

$$d_{jk} \equiv \left(\frac{2\epsilon}{M}\right) a_{jk} \Longrightarrow (\det A)^{-\frac{1}{2}} = \left(\frac{2\epsilon}{M}\right)^{\frac{m}{2}} (\det D)^{-\frac{1}{2}}\,. \tag{1.144}$$

This allows us to cast Eq.(1.142) as

$$\begin{aligned} K\left(q_f, t_f; q_i, t_i\right) &= \lim_{m\to\infty} \left(\frac{M}{2\pi\hbar i\epsilon}\right)^{\frac{1}{2}} (\det D)^{-\frac{1}{2}} \\ &\times e^{\frac{i}{\hbar}\left[R(\alpha)\,q_f^2 + H(\alpha)\,q_i^2 - \frac{c^2}{4}\right]}\,. \end{aligned} \tag{1.145}$$

---

[6] The computation of the residual integral in Eq.(A.48) is straightforward because, presently,

$$F\left(\frac{i}{\hbar}v\right) = e^{\frac{i}{\hbar}v}\,.$$

So far, all integrals entering the right hand side of Eq.(1.135) have been computed. What remains to be done is to rewrite the corresponding result, on the right hand side of Eq.(1.145), regarding the problem's data.

To start with, we shall be needing the explicit form of the matrix element d_{jk}. It can be obtained by combining Eqs.(1.144), (1.141) and (1.139). With regard to

$$r(\alpha) \equiv \left[1 - \epsilon^2 \omega^2 \left(\frac{1}{2} + \alpha\right)^2\right] , \tag{1.146a}$$

$$h(\alpha) \equiv \left[1 - \epsilon^2 \omega^2 \left(\frac{1}{2} - \alpha\right)^2\right] , \tag{1.146b}$$

$$g(\alpha) \equiv \left[-1 - \epsilon^2 \omega^2 \left(\frac{1}{4} - \alpha^2\right)\right] , \tag{1.146c}$$

we find the following

$$d_{jk} \equiv [r(\alpha) + h(\alpha)] \, \delta_{jk} + g(\alpha) \, (\delta_{j-1,k} + \delta_{j,k-1}) = d_{kj} \, . \tag{1.147}$$

We also introduce

$$u_{jk} \equiv - \frac{1}{g(\alpha)} \, d_{jk} = u(\alpha) \, \delta_{jk} - \delta_{j-1,k} - \delta_{j,k-1} \, , \tag{1.148}$$

with

$$u(\alpha) \equiv - \frac{r(\alpha) + h(\alpha)}{g(\alpha)} \, . \tag{1.149}$$

These are the elements of the matrix

$$\|U\| = \begin{bmatrix} u & -1 & 0 & 0 \ldots & 0 & 0 & 0 & 0 \\ -1 & u & -1 & 0 \ldots & 0 & 0 & 0 & 0 \\ 0 & -1 & u & -1 \ldots & 0 & 0 & 0 & 0 \\ \vdots & \vdots & \vdots & \vdots \ldots & \vdots & \vdots & \vdots & \vdots \\ 0 & 0 & 0 & 0 \ldots & -1 & u & -1 & 0 \\ 0 & 0 & 0 & 0 \ldots & 0 & -1 & u & -1 \\ 0 & 0 & 0 & 0 \ldots & 0 & 0 & -1 & u \end{bmatrix} . \tag{1.150}$$

It is explicitly displayed because the relevant outcomes from our calculations will be expressed either in terms of the determinant of $\|U\|$ and/or of its co-factors. For instance, from Eq.(1.148) it follows that

$$\det D = [-g(\alpha)]^m \det U \, . \tag{1.151}$$

On the other hand, the determination of c^2 (see Eq.(1.143)) demands the knowledge of the matrix elements g_{jk}. From Eqs.(1.144), (1.147) and (1.148) we obtain

$$\begin{aligned} g_{jk} &= (-1)^{j+k} \frac{(\text{Minor of } a_{kj})}{\det A} = (-1)^{j+k} \frac{2\epsilon}{M} \frac{(\text{Minor of } d_{kj})}{\det D} \\ &= (-1)^{j+k-1} \frac{2\epsilon}{M} g^{-1}(\alpha) \frac{(\text{Minor of } u_{kj})}{\det U} . \end{aligned} \tag{1.152}$$

Then, after some algebra we find

$$\begin{aligned} &\frac{i}{\hbar} \left[R(\alpha)\, q_f^2 + H(\alpha)\, q_i^2 - \frac{c^2}{4} \right] \\ &= \frac{i}{\hbar} \frac{M}{2\epsilon \det U} \Big\{ \left[h(\alpha) \det U + g(\alpha) \, (\text{Minor of } u_{11}) \right] q_i^2 \\ &+ \left[r(\alpha) \det U + g(\alpha) \, (\text{Minor of} \, u_{11}) \right] q_f^2 \\ &- 2(-1)^m g(\alpha) \, (\text{Minor of } u_{m1}) \, q_i q_f \Big\} . \end{aligned} \tag{1.153}$$

We now address the problem of finding $\det \|U\|$ and its co-factors. We denote by $U_k^m(u)$ the principal minor of order $m - k + 1$ in the lower right hand corner of the matrix in Eq.(1.150). Clearly

$$U_1^m(u) = \det U \,, \tag{1.154a}$$

$$U_m^m(u) = u \,. \tag{1.154b}$$

Furthermore, the development of $U_k^m(u)$ regarding the elements of the first row yields

$$U_k^m(u) - u\, U_{k+1}^m(u) - U_{k+2}^m(u) = 0 \,, \tag{1.155}$$

which is just the recursion relation defining the Tchebychev polynomials of the second class $\mathcal{U}_r(\frac{u}{2})$ $(r \equiv m - k)$ obeying the boundary condition $\mathcal{U}_1(\frac{u}{2}) = u$ [Gradshteyn and Ryzhik (1980)]. Thus, we can write

$$U_{k+1}^m(u) = \mathcal{U}_r \left(\frac{u}{2} \right) . \tag{1.156}$$

In accordance with [Gradshteyn and Ryzhik (1980)] these polynomials can be realized in terms of trigonometric functions as

$$\mathcal{U}_r \left(\frac{u}{2} \right) = \frac{\sin\left[(r+1)\varphi \right]}{\sin \varphi} . \tag{1.157}$$

Here,

$$\varphi = \arccos\frac{u}{2} = \arccos\left[\frac{1 - \frac{\epsilon^2\omega^2}{2}\left(\frac{1}{2} + 2\alpha^2\right)}{1 + \frac{\epsilon^2\omega^2}{2}\left(\frac{1}{2} - 2\alpha^2\right)}\right], \tag{1.158}$$

where Eqs.(1.146) and (1.149) have been taken into account. Therefore,

$$\det U = U_1^m(u) = \mathcal{U}_m\left(\frac{u}{2}\right) = \frac{\sin\left[(m+1)\varphi\right]}{\sin\varphi}. \tag{1.159}$$

What comes next is the evaluation of the above quantities at the limit $m \to \infty \Longrightarrow \epsilon \to 0$. Specifically,

$$\begin{aligned}\sin\varphi &= \left(1 - \cos^2\varphi\right)^{\frac{1}{2}} = \left(1 - \frac{u^2}{4}\right)^{\frac{1}{2}} \\ &= \epsilon\,\omega\,\frac{\left(1 - \epsilon^2\,\omega^2\,\alpha^2\right)^{\frac{1}{2}}}{1 + \epsilon^2\,\omega^2\left(\frac{1}{4} - \alpha^2\right)},\end{aligned} \tag{1.160}$$

implying that

$$\lim_{m\to\infty}\sin\varphi \to \epsilon\,\omega + \mathcal{O}(\epsilon^2) \tag{1.161}$$

and

$$\lim_{m\to\infty}\varphi \to \epsilon\,\omega + \mathcal{O}(\epsilon^2). \tag{1.162}$$

Hence, Eq.(1.159) yields

$$\lim_{m\to\infty}\epsilon\det U = \epsilon\,\frac{\sin\left[(m+1)\epsilon\omega\right]}{\epsilon\omega} = \frac{\sin\omega T}{\omega}, \tag{1.163}$$

where

$$T \equiv t_f - t_i\,. \tag{1.164}$$

On the other hand,

$$\lim_{m\to\infty} g(\alpha) \to -1\,, \tag{1.165}$$

as can be seen from Eq.(1.146c). Together with Eqs.(1.151) and (1.163) this leads to

$$\lim_{m\to\infty} \epsilon \det D = \frac{\sin \omega T}{\omega}, \tag{1.166}$$

which, in turn, implies that

$$\begin{aligned}\lim_{m\to\infty} \left(\frac{M}{2\pi\hbar i\epsilon}\right)^{\frac{1}{2}} (\det D)^{-\frac{1}{2}} &= \left(\frac{M}{2\pi\hbar i\epsilon \det D}\right)^{\frac{1}{2}} \\ &= \left[\frac{M\,\omega}{2\pi\hbar i\, \sin \omega T}\right]^{\frac{1}{2}}.\end{aligned} \tag{1.167}$$

We now concentrate on computing the first bracket in the right hand side of Eq.(1.153). The first piece of information comes from

$$\text{Minor of } u_{11} = U_2^m(u) = \mathcal{U}_{m-1}\left(\frac{u}{2}\right) = \frac{\sin(m\varphi)}{\sin\varphi}. \tag{1.168}$$

Thus, from Eqs.(1.146b) and (1.146c) it follows that

$$\begin{aligned}&\lim_{m\to\infty} [h(\alpha) \det U + g(\alpha)\,(\text{Minor of } u_{11})] \\ &= \lim_{m\to\infty} (\det U - \text{Minor of } u_{11}) \\ &= \lim_{m\to\infty} \left\{\frac{\sin[(m+1)\varphi] - \sin(m\varphi)}{\sin\varphi}\right\},\end{aligned} \tag{1.169}$$

which by way of the following trigonometric identity

$$\sin m\varphi = \sin[(m+1)\varphi]\cos\varphi - \cos[(m+1)\varphi]\sin\varphi \tag{1.170}$$

can be rearranged as

$$\begin{aligned}&\lim_{m\to\infty} [h(\alpha) \det U + g(\alpha)\,(\text{Minor of } u_{11})] \\ &= \lim_{m\to\infty} \left\{\frac{\sin[(m+1)\varphi]}{\sin\varphi}(1-\cos\varphi) + \cos[(m+1)\varphi]\right\}.\end{aligned} \tag{1.171}$$

By invoking Eq.(1.162) we obtain

$$\lim_{m\to\infty} [h(\alpha) \det U + g(\alpha)\,(\text{Minor of } u_{11})] = \cos\omega T. \tag{1.172}$$

The computation of the second bracket in the right hand side of Eq.(1.153) goes along similar lines and yields

$$\lim_{m\to\infty}\left[r(\alpha)\,\det U\,+\,g(\alpha)\,(\text{Minor of}\,u_{11})\right]\;=\;\cos\omega T\,. \tag{1.173}$$

As for the last term in the right hand side of Eq.(1.153) its evaluation is straightforward since Eq.(1.150) implies

$$\text{Minor of }\,u_{m1}\;=\;(-1)^{m-1} \tag{1.174}$$

so that

$$\lim_{m\to\infty}\left[-\,2(-1)^m g(\alpha)\,(\text{Minor of }\,u_{m1})\right]\;=\;+1\,. \tag{1.175}$$

By returning with Eqs.(1.172), (1.173) and (1.175) into (1.153) and, subsequently, with this latter and Eq.(1.167) into Eq.(1.145) we find for the propagator of the one dimensional harmonic oscillator the following final form

$$\begin{aligned} K\,(q_f,t_f;q_i,t_i)\;&=\;\left(\frac{M\,\omega}{2\pi\hbar i\,\sin\omega T}\right)^{\frac{1}{2}} \\ &\times\exp\left\{\frac{i}{\hbar}\frac{M\omega}{2\,\sin\omega T}\left[(q_f^2+q_i^2)\cos\omega T-2q_f\,q_i\right]\right\}, \end{aligned} \tag{1.176}$$

which involves only the data of the problem [Feynman (1948); Feynman and Hibbs (1965)]. It is worth noticing that the final result in Eq.(1.176) does not depend on α. Indeed, the lack of uniqueness introduced by the GWT was washed out by the limit $m\to\infty$ in a rather straightforward way. As we shall see later on (section 1.6) this is not always the case.

The point now is that the exponent in Eq.(1.176) turns out to be the action of the one-dimensional harmonic oscillator evaluated on the classical trajectory [Feynman (1948); Feynman and Hibbs (1965)]. To understand how this come about we recall that the Lagrangian describing the classical dynamics of the one dimensional harmonic oscillator reads

$$L(q,\dot q)=\frac{1}{2}M\dot q^2-\frac{1}{2}M\omega^2 q^2 \tag{1.177}$$

while the corresponding action is given by

$$S[q]=\int_{t_i}^{t_f}dtL(q,\dot q)\,. \tag{1.178}$$

By going with Eq.(1.177) into Eq.(1.178) and subsequently performing a by part integration we find

$$S[q] = \frac{1}{2} M\, q(t)\, \dot{q}(t)\Big|_{t_i}^{t_f} - \frac{M}{2} \int_{t_i}^{t_f} dt q(t) \left[\ddot{q}(t) + \omega^2 q(t)\right] . \quad (1.179)$$

For $q(t) = q_{cl}(t)$ the last expression is reduced to

$$S[q_{cl}] = \frac{1}{2} M\, q_{cl}(t)\, \dot{q}_{cl}(t)\Big|_{t_i}^{t_f} , \quad (1.180)$$

since $q_{cl}(t)$ solves the classical equation of motion $\ddot{q}_{cl}(t) + \omega^2 q(t)_{cl} = 0$. The analytic form of the trajectory is

$$q_{cl}(t) = A \sin \omega t + B \cos \omega t , \quad (1.181)$$

where

$$A = \frac{q_f \cos \omega t_i - q_i \cos \omega t_f}{\sin \omega T} , \quad (1.182a)$$

$$B = \frac{q_i \sin \omega t_f - q_f \sin \omega t_i}{\sin \omega T} . \quad (1.182b)$$

By replacing Eqs.(1.182) into Eq.(1.181) and, subsequently, this latter into Eq.(1.180), we find

$$S[q_{cl}] = \frac{M\omega}{2 \sin \omega T} \left[(q_f^2 + q_i^2) \cos \omega T - 2 q_f\, q_i\right] . \quad (1.183)$$

Hence, Eq.(1.176) can be rewritten as

$$K\left(q_f, t_f; q_i, t_i\right) = \left[\frac{M\,\omega}{2\pi\hbar i\, \sin \omega T}\right]^{\frac{1}{2}} e^{\frac{i}{\hbar} S[q_{cl}]} . \quad (1.184)$$

The fact that the propagator becomes *fully* determined by the classical trajectory holds true for the harmonic oscillator as well as for other exactly solvable models. On general grounds, the propagator is contributed by all *configurations* verifying the boundary conditions.

We shall furthermore prove that the expression in Eq.(1.176) verifies Eq.(1.47) or, which amounts to the same thing, that

$$\lim_{T \to 0} \left(\frac{M\,\omega}{2\pi\hbar i \sin \omega T}\right)^{\frac{1}{2}} \times \exp\left\{\frac{i}{\hbar} \frac{M\omega}{2 \sin \omega T} \left[(q_f^2 + q_i^2) \cos \omega T - 2 q_f\, q_i\right]\right\} \quad (1.185)$$

furnishes a faithful representation for the generalized function $\delta(q_f - q_i)$. To this end, we start by setting $\sin\omega T \approx \omega T$ and $\cos\omega T \approx 1$ in the right hand side of Eq.(1.185); these are correct up to terms of the first order in ωT. Thus, our task reduces to verify that

$$f(x) \equiv \left(\frac{M}{2\pi\hbar i T}\right)^{\frac{1}{2}} \exp\left\{\frac{i}{\hbar}\frac{M}{2T}x^2\right\} \tag{1.186}$$

fulfils [7]

$$\lim_{T\to 0}\int_{-\infty}^{+\infty} dx\, f(x) = 1\,, \tag{1.187a}$$

$$\lim_{T\to 0}\int_{-\infty}^{+\infty} dx\, f(x)\,\rho(x) = \rho(0)\,. \tag{1.187b}$$

Here, $\rho(x)$ is a test function defined in $\mathbb{R}_1$. The integral in Eq.(1.187a) can be performed with the help of Eq.(A.8). As it can be seen, all T factors cancel out and the result is exactly that in Eq.(1.187a). As for the integral in Eq.(1.187b), we take advantage of the analyticity of the test function to write

$$\int_{-\infty}^{+\infty} dx\, f(x)\,\rho(x) = \sum_{k=0}^{\infty}\frac{\rho^{(k)}(0)}{k!}\int_{-\infty}^{+\infty} dx\, f(x)\, x^k\,. \tag{1.188}$$

However,

$$\int_{-\infty}^{+\infty} dx\, f(x)\, x^k = \left(\frac{M}{2\pi\hbar i T}\right)^{\frac{1}{2}} \times \left(\frac{\hbar}{i}\right)^k \frac{d^k}{db^k}\int_{-\infty}^{+\infty} dx\, \exp\left\{\frac{i}{\hbar}\frac{M}{2T}x^2 + \frac{i}{\hbar}\,b\,x\right\}\bigg|_{b=0}\,, \tag{1.189}$$

which can be explicitly evaluated by using Eq.(A.10). We then find

$$\int_{-\infty}^{+\infty} dx\, f(x)\, x^k = \left(\frac{\hbar}{i}\right)^k \frac{d^k}{db^k}\, e^{-\frac{i}{\hbar}\frac{b^2 T}{2M}}\bigg|_{b=0}\,. \tag{1.190}$$

Only the term $k = 0$ survives at the limit $T \to 0$ implying that Eq.(1.188) reduces to Eq.(1.187b). As required, the expression obtained for the propagator verifies the boundary condition in Eq.(1.47).

[7] Clearly, x is short for $q_f - q_i$.

1.4.2 *Eigenvalues and eigenvectors*

Energy eigenvalues and eigenvectors are essential pieces of information in quantum mechanics. In this subsection, we shall illustrate their recovery from the propagator. We remain focused on the one dimensional harmonic oscillator.

A simple calculation shows that the main result in subsection 1.4.1,

$$\langle q_f|\, e^{-\frac{i}{\hbar}\, H(Q,P)T}\, |q_i\rangle \;=\; K(q_f, t_f; q_i, t_i) \;=\; \left(\frac{M\,\omega}{2\pi\hbar i\, \sin\omega T}\right)^{\frac{1}{2}}$$
$$\times \exp\left\{\frac{i}{\hbar}\frac{M\omega}{2\,\sin\omega T}\left[(q_f^2 + q_i^2)\cos\omega T - 2q_f\, q_i\right]\right\}, \tag{1.191}$$

leads to

$$\int_{-\infty}^{+\infty} dq_f\, \langle q_f|\, e^{-\frac{i}{\hbar}\, H(Q,P)T}\, |q_f\rangle \;=\; -\frac{i}{2\,\sin\frac{\omega T}{2}}\,. \tag{1.192}$$

As for the structure of the energy eigenvalue spectrum we begin by assuming that it is made up of a discrete and continuous part which does not overlap. This enables us to write the spectral resolution of the identity operator as

$$I = \sum_n |E_n\rangle\langle E_n| + \int_0^{+\infty} dE\,|E\rangle\langle E|\,. \tag{1.193}$$

By inserting Eq.(1.193) into the left hand side of Eq.(1.192) we find

$$\sum_n e^{-\frac{i}{\hbar}E_n\, T} \int_{-\infty}^{+\infty} dq_f\, \langle E_n|q_f\rangle\langle q_f|E_n\rangle$$
$$+ \int_0^{+\infty} dE\, e^{-\frac{i}{\hbar}E\,T} \int_{-\infty}^{+\infty} dq_f\, \langle E|q_f\rangle\langle q_f|E\rangle$$
$$= -\frac{i}{2\,\sin\frac{\omega T}{2}}\,, \tag{1.194}$$

which, on account of Eq.(1.34), is equivalent to

$$\sum_n e^{-\frac{i}{\hbar}E_n\, T}\langle E_n|E_n\rangle + \int_0^{+\infty} dE\, e^{-\frac{i}{\hbar}E\,T}\langle E|E\rangle$$
$$= -\frac{i}{2\,\sin\frac{\omega T}{2}}\,. \tag{1.195}$$

We look next for the elimination of the exponential factors in the left hand side of this last equation. To that end, we shall continue T to the imaginary axis and then introduce the real Euclidean time τ by means of

$$T \longrightarrow -i\,\tau\,. \tag{1.196}$$

Accordingly, Eq.(1.195) goes into

$$\sum_n \langle E_n|E_n\rangle\, e^{-\frac{E_n}{\hbar}\,\tau} + \int_0^{+\infty} dE\,\langle E|E\rangle\, e^{-\frac{E}{\hbar}\,\tau} = \frac{1}{2\sinh\frac{\omega\tau}{2}}\,. \tag{1.197}$$

We now multiply both sides of Eq.(1.197) by τ and integrate on this variable from 0 to $+\infty$. The result is

$$\sum_n \langle E_n|E_n\rangle \left(\frac{E_n}{\hbar}\right)^{-2} + \int_0^{+\infty} dE\,\langle E|E\rangle \left(\frac{E}{\hbar}\right)^{-2} = \frac{1}{2}\left(\frac{\pi}{\omega}\right)^2\,, \tag{1.198}$$

where we have used [Gradshteyn and Ryzhik (1980)]

$$\int_0^{+\infty} d\tau\,\tau\,e^{-\frac{E}{\hbar}\,\tau} = \left(\frac{E}{\hbar}\right)^{-2} \tag{1.199}$$

and

$$\int_0^{+\infty} d\tau\,\frac{\tau}{\sinh\frac{\omega\tau}{2}} = \left(\frac{\pi}{\omega}\right)^2\,. \tag{1.200}$$

The energy eigenvectors describing scattering states are of infinite norm while the right hand side of Eq.(1.198) is finite. This rules out the possibility for the existence of a continuous spectrum. Consequently, Eq.(1.193) is simplified as follows

$$I = \sum_{n=0}^{\infty} |E_n\rangle\langle E_n|\,. \tag{1.201}$$

As for the eigenvectors of the discrete spectrum one does not lose generality by assuming that they are orthonormal. Therefore, Eq.(1.198) reduces to

$$\sum_n \left(\frac{E_n}{\hbar}\right)^{-2} = \frac{1}{2}\left(\frac{\pi}{\omega}\right)^2\,. \tag{1.202}$$

Dimensional analysis alone signalizes that the energy eigenvalues must be of the following form

$$E_n = \hbar\omega F_n , \tag{1.203}$$

where F_n is a dimensionless quantity to be determined from

$$\sum_n \left(\frac{1}{F_n^2}\right) = \frac{\pi^2}{2} . \tag{1.204}$$

From [Gradshteyn and Ryzhik (1980)] we get

$$\sum_{n=0}^{\infty} \frac{1}{(2n+1)^2} = \frac{\pi^2}{8} , \tag{1.205}$$

yielding

$$F_n = n + \frac{1}{2} , \qquad n = 0, \ldots , \tag{1.206}$$

which along with Eq.(1.203) leads to

$$E_n = \left(n + \frac{1}{2}\right) \hbar\omega , \tag{1.207}$$

in agreement with the outcome from the operator approach.

What comes next is the problem of determining the eigenfunctions ($\psi_n(q) \equiv \langle q|E_n\rangle$) associated with the eigenvalues in Eq.(1.207). They explicitly emerge as Eq.(1.191) is rewritten as

$$\sum_{n=0}^{\infty} \psi_n(q_f)\, e^{-i\,(n+\frac{1}{2})\,\omega T}\, \psi_n(q_i) = \left(\frac{M\omega}{2\pi\hbar i\, \sin\omega T}\right)^{\frac{1}{2}}$$
$$\times \exp\left\{\frac{i}{\hbar}\frac{M\omega}{2\,\sin\omega T}\left[(q_f^2 + q_i^2)\cos\omega T - 2q_f\, q_i\right]\right\} , \tag{1.208}$$

where Eq.(1.207) has been taken into account. The exponential in the right hand side of Eq.(1.208) reminds us of the generating function of the Hermite polynomials ($H_n(x)$). In accordance with [Erdélyi *et al.* (1953)]

$$\sum_{n=0}^{\infty} \frac{z^n}{2^n\, n!} H_n(x_f)\, H_n(x_i)$$
$$= (1-z^2)^{-\frac{1}{2}} \exp\left[\frac{2zx_f x_i - z^2\left(x_f^2 + x_i^2\right)}{1-z^2}\right] . \tag{1.209}$$

Here, the real variables x_f and x_i as well as the complex variable z are dimensionless. For

$$z = e^{-i\omega T} \tag{1.210}$$

Eq.(1.209) becomes

$$\sum_{n=0}^{\infty} \frac{1}{2^n\, n!} H_n(x_f)\, e^{-in\omega T}\, H_n(x_i) = e^{\frac{i\omega T}{2}} \frac{1}{(2i \sin \omega T)^{\frac{1}{2}}}$$
$$e^{\frac{1}{2}(x_f^2+x_i^2)} \exp\left\{ \frac{i}{2\sin\omega T} \left[(x_f^2 + x_i^2) \cos\omega T - 2x_f x_i \right] \right\} \tag{1.211}$$

which, after algebraic rearrangements, can be written as

$$\sum_{n=0}^{\infty} \left\{ \left[\left(\frac{1}{2^n\, n!} \right)^{\frac{1}{2}} e^{-\frac{1}{2}x_f^2} H_n(x_f) \right] e^{-i(n+\frac{1}{2})\omega T} \left[\left(\frac{1}{2^n\, n!} \right)^{\frac{1}{2}} e^{-\frac{1}{2}x_i^2} H_n(x_i) \right] \right\}$$
$$= \frac{1}{(2i\sin\omega T)^{\frac{1}{2}}} \exp\left\{ \frac{i}{2\sin\omega T} \left[(x_f^2 + x_i^2)\cos\omega T - 2x_f x_i \right] \right\}. \tag{1.212}$$

However, we are required to write the dimensionless variable x in terms of the variable q with dimensions of length (cm^1). The only dimensional constants at hand are M, $\hbar$ and ω. So, we must find α, β and γ such that

$$x = M^{\alpha}\, \hbar^{\beta}\, \omega^{\gamma}\, q\,. \tag{1.213}$$

We find that $\alpha = -\beta = \gamma = +1/2$ is the unique solution. Therefore,

$$x = \left(\frac{M\omega}{\hbar} \right)^{\frac{1}{2}} q\,. \tag{1.214}$$

Hence, Eq.(1.212) can be cast

$$\sum_{n=0}^{\infty} \left\{ \left[\left(\frac{1}{2^n\, n!} \right)^{\frac{1}{2}} e^{-\frac{M\omega}{2\hbar} q_f^2} H_n\left(\sqrt{\frac{M\omega}{\hbar}} q_f \right) \right] e^{-i(n+\frac{1}{2})\omega T} \right.$$
$$\left. \times \left[\left(\frac{1}{2^n\, n!} \right)^{\frac{1}{2}} e^{-\frac{M\omega}{2\hbar} q_i^2} H_n\left(\sqrt{\frac{M\omega}{\hbar}} q_i \right) \right] \right\}$$
$$= \frac{1}{(2i\sin\omega T)^{\frac{1}{2}}} \exp\left\{ \frac{iM\omega}{2\hbar\sin\omega T} \left[(q_f^2 + q_i^2)\cos\omega T - 2q_f q_i \right] \right\} \tag{1.215}$$

or, after multiplying both sides by $\left(\frac{M\omega}{\pi\hbar}\right)^{\frac{1}{2}}$,

$$\sum_{n=0}^{\infty}\left\{\left[\left(\frac{1}{2^n\,n!}\right)^{\frac{1}{2}}\left(\frac{M\omega}{\pi\hbar}\right)^{\frac{1}{4}} e^{-\frac{M\omega}{2\hbar}q_f^2} H_n\left(\sqrt{\frac{M\omega}{\hbar}}q_f\right)\right] e^{-i\left(n+\frac{1}{2}\right)\omega T}\right.$$
$$\left.\times\left[\left(\frac{1}{2^n\,n!}\right)^{\frac{1}{2}}\left(\frac{M\omega}{\pi\hbar}\right)^{\frac{1}{4}} e^{-\frac{M\omega}{2\hbar}q_i^2} H_n\left(\sqrt{\frac{M\omega}{\hbar}}q_i\right)\right]\right\} = \left(\frac{M\omega}{2\pi\hbar i\sin\omega T}\right)^{\frac{1}{2}}$$
$$\times\exp\left\{\frac{iM\omega}{2\hbar\sin\omega T}\left[\left(q_f^2+q_i^2\right)\cos\omega T - 2q_f q_i\right]\right\}. \tag{1.216}$$

By combining Eqs.(1.208) and (1.216) we arrive at

$$\sum_{n=0}^{\infty}\psi_n(q_f)\, e^{-i\left(n+\frac{1}{2}\right)\omega T}\,\psi_n(q_i)$$
$$= \sum_{n=0}^{\infty}\left\{\left[\left(\frac{1}{2^n\,n!}\right)^{\frac{1}{2}}\left(\frac{M\omega}{\pi\hbar}\right)^{\frac{1}{4}} e^{-\frac{M\omega}{2\hbar}q_f^2} H_n\left(\sqrt{\frac{M\omega}{\hbar}}q_f\right)\right] e^{-i\left(n+\frac{1}{2}\right)\omega T}\right.$$
$$\left.\times\left[\left(\frac{1}{2^n\,n!}\right)^{\frac{1}{2}}\left(\frac{M\omega}{\pi\hbar}\right)^{\frac{1}{4}} e^{-\frac{M\omega}{2\hbar}q_i^2} H_n\left(\sqrt{\frac{M\omega}{\hbar}}q_i\right)\right]\right\}, \tag{1.217}$$

from which the eigenfunctions we are looking for are found to be

$$\psi_n(q) = \left(\frac{1}{2^n\,n!}\right)^{\frac{1}{2}}\left(\frac{M\omega}{\pi\hbar}\right)^{\frac{1}{4}}$$
$$\times\, e^{-\frac{M\omega}{2\hbar}q^2} H_n\left(\sqrt{\frac{M\omega}{\hbar}}q\right), n = 0, 1, \ldots. \tag{1.218}$$

They obey the normalization condition

$$\int_{-\infty}^{+\infty} dq\,|\psi_n(q)|^2 = 1\,. \tag{1.219}$$

The recovery of the energy spectrum from the propagator is now complete.

1.5 One dimensional free particle

Next, we shall test the systematics presented in subsection 1.4.2 in connection with a system possessing a continuous energy spectrum. The one dimensional free particle of mass M is the simplest candidate for that purpose.

The Hamiltonian operator reads

$$H(Q,P) = \frac{P^2}{2M} \tag{1.220}$$

while the corresponding propagator,

$$K(q_f,t_f;q_i,t_i) = \left(\frac{M}{2\pi\hbar i T}\right)^{\frac{1}{2}} \exp\left[\frac{i}{\hbar}\frac{M}{2T}(q_f-q_i)^2\right], \tag{1.221}$$

is obtained by setting ω to zero in Eq.(1.176). The exponent in this latter is, as expected, the free particle action evaluated on the classical trajectory.

One should also observe that Eq.(1.221) is formally identical to Eq.(1.186). This secures that the one dimensional free particle propagator fulfils, as required, the boundary condition in Eq.(1.47). As for the energy spectrum we first notice that at the limit $\omega \to 0$ Eq.(1.191) reduces to

$$\langle q_f| e^{-\frac{i}{\hbar}HT} |q_i\rangle = \left(\frac{M}{2\pi\hbar i T}\right)^{\frac{1}{2}} \exp\left[\frac{i}{\hbar}\frac{M}{2T}(q_f-q_i)^2\right], \tag{1.222}$$

which in view of Eq.(A.10) can be cast

$$\langle q_f|e^{-\frac{i}{\hbar}HT}|q_i\rangle = \frac{1}{2\pi\hbar}\int_{-\infty}^{+\infty} dp \exp\left[-\frac{i}{\hbar}\frac{p^2}{2M}T + \frac{i}{\hbar}p(q_f-q_i)\right]. \tag{1.223}$$

By introducing in the left hand side of Eq.(1.223) the identity operator written in terms of the eigenvectors of the linear momentum operator (see Eq.(1.43)) we achieve

$$\begin{aligned}&\int_{-\infty}^{+\infty} dp\langle q_f|p\rangle e^{-\frac{i}{\hbar}\frac{p^2}{2M}T}\langle p|q_i\rangle \\ &= \frac{1}{2\pi\hbar}\int_{-\infty}^{+\infty} dp \exp\left[-\frac{i}{\hbar}\frac{p^2}{2M}T + \frac{i}{\hbar}p(q_f-q_i)\right].\end{aligned} \tag{1.224}$$

Thus, the energy eigenvalues,

$$E = \frac{p^2}{2M}, \tag{1.225}$$

run continuously from 0 to $+\infty$ whereas the energy eigenfunctions are

$$\psi_p(q) \equiv \langle q|p\rangle = \frac{1}{(2\pi\hbar)^{\frac{1}{2}}}\, e^{\frac{i}{\hbar}qp}\,. \tag{1.226}$$

They obey the following normalization condition

$$\int_{-\infty}^{+\infty} d^3q\, \psi_p^*(q)\, \psi_{p'}(q) = \delta(p-p')\,. \tag{1.227}$$

These findings corroborate with the corresponding results arising from the operator framework.

1.6 The stochastic nature of the path integral

Under the canonical transformation

$$Q \longrightarrow \frac{1}{\sqrt{2}}\left(Q - \frac{P}{\omega M}\right), \tag{1.228a}$$

$$P \longrightarrow \sqrt{2}\,P\,, \tag{1.228b}$$

one can verify that the Hamiltonian operator

$$\bar{H} = \frac{P^2}{2M} + M\omega^2 Q^2 + \frac{\omega}{2}\,(QP + PQ) \tag{1.229}$$

maps onto the one describing the dynamics of the harmonic oscillator (denoted by H and given in Eq.(1.133)). Henceforth, $\bar{H}$ will be referred to as the *modified harmonic oscillator* Hamiltonian operator [Simões (1980)].

Although $\bar{H}$ and H give rise to the *same physics* the phase space path integral formulations deriving from these Hamiltonian operators differ, mainly in connection with the structure of the α-dependent terms. To begin with, the GWT of H (denoted by h_α and given in Eq.(1.134)) does not depend on α while that of $\bar{H}$,

$$\bar{h}_\alpha = \frac{p^2}{2M} + M\omega^2 q^2 + \omega q p - i\,\hbar\,\omega\,\alpha\,, \tag{1.230}$$

contains an α-dependent term which, as we shall see, *does not vanish* at the limit $\epsilon \longrightarrow 0$. Since the *same physics* means the *same propagator* we are forced to conclude that an α-dependence cancelation mechanism should take place in order to secure the consistency of the formulation. Identifying this mechanism and understanding how it works are our purposes in this section.

In accordance with Eqs.(1.128) and (1.230) the propagator for the one dimensional modified harmonic oscillator is given by

$$\bar{K}(q_f, t_f; q_i, t_i) = \lim_{m\to\infty} (2\pi\hbar)^{-(m+1)} \int_{-\infty}^{+\infty} \left(\prod_{j=1}^{m} dq_j\right) \int_{-\infty}^{+\infty} \left(\prod_{j=0}^{m} dp_j\right)$$
$$\times \exp\left\{\frac{i}{\hbar}\sum_{j=0}^{m}\epsilon\left[p_j\frac{(q_{j+1}-q_j)}{\epsilon}\right.\right.$$
$$\left.\left.-\frac{p_j^2}{2M} - M\omega^2 q_j^2(\alpha) - \omega p_j q_j(\alpha) + i\hbar\omega\alpha\right]\right\}, \tag{1.231}$$

which after carrying out the momentum integrations becomes

$$\bar{K}(q_f, t_f; q_i, t_i)$$
$$= \lim_{m\to\infty}\left[\left(\frac{M}{2\pi i\hbar\epsilon}\right)^{\frac{(m+1)}{2}} e^{-\alpha\omega(m+1)\epsilon}\,\mathcal{O}(m,\,\epsilon,\,q_f,\,q_i,\,M)\right], \tag{1.232}$$

where

$$\mathcal{O}(m,\,\epsilon,\,q_f,\,q_i,\,M) \equiv \int_{-\infty}^{+\infty}\left(\prod_{j=1}^{m} dq_j\right) e^{\frac{i}{\hbar}E(m,\,\epsilon,\,q_f,\,q,\,q_i,\,M)} \tag{1.233}$$

and

$$E(m,\,\epsilon,\,q_f,\,q,\,q_i,\,M)$$
$$\equiv \sum_{j=0}^{m}\left\{\frac{M}{2\epsilon}\left[(q_{j+1}-q_j) - \epsilon\omega q_j(\alpha)\right]^2 - M\omega^2\epsilon q_j^2(\alpha)\right\}. \tag{1.234}$$

The α-dependent term

$$e^{-\alpha\omega(m+1)\epsilon} \tag{1.235}$$

in Eq.(1.232), deserves special attention. Its origin can be traced back to the last term in the right hand side of Eq.(1.230). Above all, it survives at the limit $m \longrightarrow \infty(\epsilon \longrightarrow 0)$ since

$$\lim_{m\to\infty} e^{-\alpha\omega(m+1)\epsilon} = e^{-\alpha\omega T} \neq 0\,. \tag{1.236}$$

Consistency (α-independence) calls for its cancellation. Only the terms in $\mathcal{O}$ bilinear in $\epsilon\alpha$ can do this job. In order to see whether this happens or not we shall start by analyzing the ϵ expansion of a generic term in E, namely,

$$\begin{aligned}&\frac{M}{2\epsilon}\left[(q_{j+1}-q_j)-\epsilon\omega q_j\alpha\right]^2 - M\omega^2\epsilon q_j^2\alpha\\&=\epsilon\left(\frac{M}{2}\frac{\Delta_j^2}{\epsilon^2}-\frac{M\omega^2\tilde{q}_j^2}{2}-M\omega\tilde{q}_j\frac{\Delta_j}{\epsilon}\right)\\&-\epsilon^2\left(M\omega\alpha\frac{\Delta_j^2}{\epsilon^2}+M\omega^2\alpha\tilde{q}_j\frac{\Delta_j}{\epsilon}\right)-\epsilon^3\frac{M\omega^2\alpha^2}{2}\frac{\Delta_j^2}{\epsilon^2}\,,\end{aligned}\tag{1.237}$$

where

$$\Delta_j \equiv q_{j+1}-q_j \tag{1.238}$$

and

$$\tilde{q}_j \equiv \frac{q_{j+1}+q_j}{2}\,. \tag{1.239}$$

Clearly, E *does not* contain terms bilinear in $\epsilon\alpha$. However, this is not the end of the story because what really matters is whether or not that kind of term shows up in $\mathcal{O}$. In fact, the functional dependence of $\mathcal{O}$ on ϵ is not that of E. To derive definite conclusions we must first carry out the q-integrals in Eq.(1.233). That is our next task.

For pedagogical and technical reasons we shall parallel the calculations in subsection 1.4.1 as much as possible. To begin with

$$\begin{aligned}&\frac{M}{2\epsilon}\left[(q_{j+1}-q_j)-\epsilon\omega q_j\alpha\right]^2 - M\omega^2\epsilon q_j^2\alpha\\&=\bar{R}(\alpha)\,q_f^2+\bar{H}(\alpha)\,q_i^2+\sum_{j=1}^{m}\bar{t}_j\,q_j+\sum_{j,k=1}^{m}q_j\,\bar{a}_{jk}\,q_k\,,\end{aligned}\tag{1.240}$$

where

$$\bar{R}(\alpha)\equiv\frac{M}{2\epsilon}\left[1-2\epsilon\omega\left(\frac{1}{2}+\alpha\right)-\epsilon^2\omega^2\left(\frac{1}{2}+\alpha\right)^2\right],\tag{1.241a}$$

$$\bar{H}(\alpha)\equiv\frac{M}{2\epsilon}\left[1+2\epsilon\omega\left(\frac{1}{2}-\alpha\right)-\epsilon^2\omega^2\left(\frac{1}{2}-\alpha\right)^2\right],\tag{1.241b}$$

$$\bar{G}(\alpha)\equiv\frac{M}{2\epsilon}\left[-1+2\epsilon\omega\alpha-\epsilon^2\omega^2\left(\frac{1}{4}-\alpha^2\right)\right],\tag{1.241c}$$

whereas

$$\bar{t}_1 \equiv 2\,\bar{G}(\alpha)\,q_i\,, \quad \bar{t}_2 = \ldots = \bar{t}_{m-1} \equiv 0\,, \quad \bar{t}_m \equiv 2\,\bar{G}(\alpha)\,q_f\,, \tag{1.242}$$

and

$$\bar{a}_{jk} \equiv \left[\bar{R}(\alpha) + \bar{H}(\alpha)\right]\,\delta_{jk} + \bar{G}(\alpha)\,\left(\delta_{j-1,k} + \delta_{j,k-1}\right) = \bar{a}_{kj}\,. \tag{1.243}$$

From Eqs.(1.240), (1.234), (1.233) and (1.232) as well as after using Eq.(A.24) we find

$$\begin{aligned}&\bar{K}\left(q_f, t_f; q_i, t_i\right)\\ &= \lim_{m\to\infty}\left(\frac{M}{2\pi\hbar i\epsilon}\right)^{\frac{1}{2}} e^{-\alpha\omega(m+1)\epsilon}\left(\det\bar{D}\right)^{-\frac{1}{2}}\\ &\times e^{\frac{i}{\hbar}\left[\bar{R}(\alpha)\,q_f^2 + \bar{H}(\alpha)\,q_i^2 - \frac{\bar{c}^2}{4}\right]}\,,\end{aligned} \tag{1.244}$$

where the elements of the matrix $\|\bar{D}\|$ are

$$\bar{d}_{jk} \equiv \left[\bar{r}(\alpha) + \bar{h}(\alpha)\right]\,\delta_{jk} + \bar{g}(\alpha)\,\left(\delta_{j-1,k} + \delta_{j,k-1}\right) = \bar{d}_{kj} \tag{1.245}$$

with

$$\bar{r}(\alpha) \equiv \left[1 - 2\epsilon\omega\left(\frac{1}{2}+\alpha\right) - \epsilon^2\omega^2\left(\frac{1}{2}+\alpha\right)^2\right], \tag{1.246a}$$

$$\bar{h}(\alpha) \equiv \left[1 + 2\epsilon\omega\left(\frac{1}{2}-\alpha\right) - \epsilon^2\omega^2\left(\frac{1}{2}-\alpha\right)^2\right], \tag{1.246b}$$

$$\bar{g}(\alpha) \equiv \left[-1 + 2\epsilon\omega\alpha - \epsilon^2\omega^2\left(\frac{1}{4}-\alpha^2\right)\right]. \tag{1.246c}$$

Furthermore

$$\bar{c}^2 = \sum_{j,\,k=1}^{m} \bar{t}_j \bar{g}_{jk} \bar{t}_k\,, \tag{1.247}$$

where $\bar{g}_{jk}$ are the matrix elements of $\|\bar{A}\|^{-1}$. The elements of $\|\bar{A}\|$ are given in Eq.(1.243).

The counterparts of Eqs.(1.148), (1.149), (1.150) and (1.151) are, respectively,

$$\bar{u}_{jk} \equiv -\frac{1}{\bar{g}(\alpha)}\,\bar{d}_{jk} = \bar{u}(\alpha)\,\delta_{jk} - \delta_{j-1,k} - \delta_{j,k-1}\,, \tag{1.248}$$

$$\bar{u}(\alpha) \equiv -\frac{\bar{r}(\alpha) + \bar{h}(\alpha)}{\bar{g}(\alpha)}, \tag{1.249}$$

$$\|\bar{U}\| = \begin{bmatrix} \bar{u} & -1 & 0 & 0 \dots & 0 & 0 & 0 & 0 \\ -1 & \bar{u} & -1 & 0 \dots & 0 & 0 & 0 & 0 \\ 0 & -1 & \bar{u} & -1 \dots & 0 & 0 & 0 & 0 \\ \vdots & \vdots & \vdots & \vdots \dots & \vdots & \vdots & \vdots & \vdots \\ 0 & 0 & 0 & 0 \dots & -1 & \bar{u} & -1 & 0 \\ 0 & 0 & 0 & 0 \dots & 0 & -1 & \bar{u} & -1 \\ 0 & 0 & 0 & 0 \dots & 0 & 0 & -1 & \bar{u} \end{bmatrix}, \tag{1.250}$$

and

$$\det \bar{D} = [-\bar{g}(\alpha)]^m \det \bar{U}. \tag{1.251}$$

Through a chain of arguments similar to that following Eq.(1.155) we arrive at

$$\det \bar{U} = \frac{\sin[(m+1)]\bar{\varphi}}{\sin \bar{\varphi}}, \tag{1.252}$$

with

$$\bar{\varphi} \equiv \arccos \frac{\bar{u}}{2} = \arccos \left[\frac{1 - 2\epsilon\omega\alpha - \frac{\epsilon^2\omega^2}{2}\left(\frac{1}{2} + 2\alpha^2\right)}{1 - 2\epsilon\omega\alpha + \frac{\epsilon^2\omega^2}{2}\left(\frac{1}{2} - 2\alpha^2\right)} \right]. \tag{1.253}$$

Therefore,

$$\lim_{m\to\infty} \bar{\varphi} \to \epsilon\omega + \mathcal{O}(\epsilon^2) \tag{1.254}$$

and

$$\lim_{m\to\infty} (\sin \bar{\varphi}) \to \epsilon\omega + \mathcal{O}(\epsilon^2). \tag{1.255}$$

Then, Eq.(1.252) yields

$$\lim_{m\to\infty} \epsilon \det \bar{U} = \epsilon \frac{\sin[(m+1)\epsilon\omega]}{\epsilon\omega} = \frac{\sin \omega T}{\omega}. \tag{1.256}$$

This corroborates with the result in Eq.(1.163). However, instead of Eq.(1.165) we now find (recalling Eq.(1.246c))

$$\lim_{m\to\infty}[-\bar{g}(\alpha)]^{-\frac{m}{2}} = \lim_{m\to\infty}(1-2\epsilon\omega\alpha) = \lim_{m\to\infty}\left(e^{-2\epsilon\omega\alpha}\right)^{-\frac{m}{2}}$$
$$= \lim_{m\to\infty} e^{\epsilon m\omega\alpha} = e^{\alpha\omega T}\,. \tag{1.257}$$

By going back with Eqs.(1.257) and (1.256) into Eq. (1.251) we obtain

$$\lim_{m\to\infty}\left(\epsilon\det\bar{D}\right)^{-\frac{1}{2}} = e^{\alpha\omega T}\left(\frac{\omega}{\sin\omega T}\right)^{\frac{1}{2}} \tag{1.258}$$

and, consequently,

$$\lim_{m\to\infty}\left(\frac{M}{2\pi\hbar i\epsilon}\right)^{\frac{1}{2}} e^{-\alpha\omega(m+1)\epsilon}\left(\det\bar{D}\right)^{-\frac{1}{2}} = \left(\frac{M\omega}{2\pi\hbar i\sin\omega T}\right)^{\frac{1}{2}}\,. \tag{1.259}$$

The contribution to $\bar{K}$ arising from the last exponential in the right hand side of Eq.(1.244) turns out to be

$$e^{\frac{i}{\hbar}\left[\bar{R}(\alpha)\,q_f^2 + \bar{H}(\alpha)\,q_i^2 - \frac{\bar{c}^2}{4}\right]} = e^{\frac{i}{\hbar}S[q_{cl}]}\,, \tag{1.260}$$

where S is the action deriving from H. By putting everything back together into Eq.(1.244) we find, as expected,

$$\bar{K}\left(q_f,t_f;q_i,t_i\right) = K\left(q_f,t_f;q_i,t_i\right)\,. \tag{1.261}$$

Therefore, the α-dependence is gone and the final result agrees with that obtained from starting with the Hamiltonian in Eq.(1.133).

The mechanism of cancellation of the α dependence works as follows. Along the computation of $\det\bar{D}$ an exponential whose exponent depends linearly on the product $\alpha\epsilon$ emerges (see Eq.(1.257)). It entirely washes out the α dependence from $\bar{K}$. This exponential is a consequence of the term linear in ϵ in Eq.(1.246c) which, in turn, originates from one of the terms of order ϵ^2 in Eq.(1.237). Roughly speaking, terms that are of order ϵ^2 before carrying out the q-integrals, become of order ϵ after such integrations are performed. This is known as the *stochastic nature of the path integral* [Gervais and Jevicki (1976); Edwards and Gulyaev (1964)]. Nevertheless, a proof insuring the uniqueness of the phase space path integral representation of the propagator, in the general case, is still lacking.

1.7 Problem

Problem 1-1

Show that

$$
\begin{aligned}
&(2\pi\hbar)^{-N}\int_{-\infty}^{+\infty} d^N\tau\, e^{-\frac{i}{\hbar}\tau\cdot p}\left|q-\left(\frac{1}{2}+\alpha\right)\tau\right\rangle\left\langle q+\left(\frac{1}{2}-\alpha\right)\tau\right| \\
&= (2\pi\hbar)^{-2N}\int_{-\infty}^{+\infty} d^N u \int_{-\infty}^{+\infty} d^N v\, e^{\frac{i}{\hbar}[(q-Q)\cdot u+(p-P)\cdot v]}\, e^{\frac{i}{\hbar}\alpha v\cdot u}\,.
\end{aligned}
$$

This result substantiates the equality in Eq.(1.51).

Chapter 2

Schwinger equations

This Chapter aims to present a unified view of the functional formulation of quantum mechanics. Fictitious sources for coordinates and momenta are introduced and the dynamics in the fictitious source picture is formulated. The Schwinger action principle is derived and used, afterwards, for introducing the Green functions. The functional differential equations verified by the Green functions generating functional, i.e., Schwinger equations, are derived and integrated. It is amusing that the functional Fourier transform solving Schwinger equations turns out to be the phase space path integral. The role played by the GWT within the framework of Schwinger equations is clarified.

We reintroduce the subscripts or superscripts S and H for labeling quantities belonging to the Schrödinger and Heisenberg pictures, respectively.

2.1 Fictitious source picture

We seek for the operator quantum dynamics of a system possessing N degrees of freedom in configuration space. We shall denote by $\{Q\} \equiv Q^1, \ldots, Q^N$ and $\{P\} \equiv P_1, \ldots, P_N$ the Cartesian position operators and their canonical conjugate momenta, respectively [1]. They verify the following equal-time canonical commutation rules

$$[Q^j(t)\,,\, Q^k(t)] = 0\,, \tag{2.1a}$$

$$[Q^j(t)\,,\, P_k(t)] = i\hbar\,\delta^j{}_k\, I\,, \tag{2.1b}$$

$$[P_j(t)\,,\, P_k(t)] = 0\,. \tag{2.1c}$$

[1]The covariant notation for coordinates and momenta will be adopted from now on.

The dynamics of the system is induced by the Hamiltonian operator

$$H\,(Q,P)\,+\,H^{(s)}\,, \tag{2.2}$$

where [2]

$$H^{(s)}\;\equiv\;-\,Q^j(t)J_j(t)\;-\;P_j(t)K^j(t)\,. \tag{2.3}$$

Here, the c-numbers $\{J\}\equiv J_1,\ldots,J_N$ and $\{K\}\equiv K^1,\ldots,K^N$ designate the *fictitious sources* of Q and P, respectively [3]. Heisenberg picture operators $(L_H(t))$ evolve in time according to the equation of motion [4]

$$\frac{dL_H(t)}{dt}\;=\;\frac{i}{\hbar}\left[H_H(t)+H_H^{(s)}(t)\,,\,L_H(t)\right]\,, \tag{2.4}$$

while the state vector $(|\Psi(t)>_H)$ remains stationary, i.e.,

$$\frac{d|\Psi(t)\rangle_H}{dt}\;=0\,. \tag{2.5}$$

We shall be seeking for a new picture where the time evolution of operators is induced by H while that of the state vector is due to $H^{(s)}$. It will be referred to as the *fictitious source picture.* We adopt the convention that quantities belonging to the fictitious source picture will bear no label. Moreover, we denote by $\tilde{U}(t,t_i)$ the unitary operator implementing the canonical transformation linking the Heisenberg with the fictitious source picture, namely,

$$L_H(t)\longrightarrow L(t)\;=\;\tilde{U}(t,t_i)\,L_H(t)\,\tilde{U}^\dagger(t,t_i)\,, \tag{2.6a}$$

$$|\Psi\rangle_H\longrightarrow|\Psi(t)\rangle\;=\;\tilde{U}(t,t_i)\,|\Psi\rangle_H\,. \tag{2.6b}$$

Both pictures coincide at $t=t_i$. Hence, the operator $\tilde{U}(t,t_i)$ must verify the boundary condition

$$\tilde{U}(t_i,t_i)\;=\;I\,. \tag{2.7}$$

[2] The superscript s, referring collectively to the *fictitious sources* should not be confused with the S denoting Schrödinger picture quantities.

[3] Alternatively, $\{J\}$ and $\{K\}$ are also referred to as *external sources.*

[4] Exception made of $H^{(s)}$ no other operator in the theory depends *explicitly* on time.

We claim that $\tilde{U}(t,t_i)$ is the solution to the following differential equation

$$\frac{d\tilde{U}(t,t_i)}{dt} = -\frac{i}{\hbar} H^{(s)}(t)\, \tilde{U}(t,t_i)\,. \tag{2.8}$$

Indeed, by differentiating both sides of Eq.(2.6b) with respect to time we find, in accordance with Eq.(2.8),

$$\frac{d|\Psi(t)\rangle}{dt} = -\, i\, \hbar\, H^{(s)}(t)\, |\Psi(t)\rangle\,, \tag{2.9}$$

while from Eqs.(2.6a) and (2.8) it follows that

$$\frac{dL(t)}{dt} = \frac{i}{\hbar}\, [H(t)\,,\, L(t)]\ . \tag{2.10}$$

As intended, $H^{(s)}$ accounts for the time development of the state vector whereas H is responsible for the evolution in time of operators. The integral equation

$$\tilde{U}[s|t_f,t_i] = I - \frac{i}{\hbar} \int_{t_i}^{t_f} dt\, H^{(s)}(t)\, \tilde{U}[s|t,t_i]\,, \tag{2.11}$$

summarizes the differential equation (2.8) and the initial condition (2.7). Notice that the *functional* dependence of $\tilde{U}$ on s has now been explicitly displayed.

At the limit of vanishing fictitious sources ($H^{(s)} = 0$) Eq.(2.11) reduces to

$$\tilde{U}[s=0|t_f,t_i] = I \tag{2.12}$$

and, consequently, the fictitious source picture coincides with the Heisenberg picture. We shall see that, nevertheless, the functional derivatives of $\tilde{U}[s|t_f,t_i]$ with respect to the fictitious sources play a relevant role at this limit.

2.2 Schwinger action principle

We now look for the change in $\tilde{U}[s|t_f,t_i]$ provoked by an infinitesimal change in the fictitious sources, i.e.,

$$J \to J + \delta J, \quad K \to K + \delta K \Longrightarrow H^{(s)} \to H^{(s)} + \delta H^{(s)} \,. \tag{2.13}$$

By assumption, the functional dependence of $\tilde{U}[s|t_f, t_i]$ on the fictitious sources is continuous. Hence, the response of $\tilde{U}[s|t_f, t_i]$ to an infinitesimal change of the fictitious sources is

$$\tilde{U}[s|t_f, t_i] \longrightarrow \tilde{U}[s + \delta s \,|\, t_f, t_i] \;=\; \tilde{U}[s|t_f, t_i] \,+\, \delta\tilde{U}[s|t_f, t_i] \,. \tag{2.14}$$

Furthermore, from Eq.(2.11) it follows that

$$\begin{aligned} \tilde{U}[s|t_f, t_i] \,+\, \delta\tilde{U}[s|t_f, t_i] \;&=\; I \\ -\frac{i}{\hbar}\int_{t_i}^{t_f} dt \, \left(H^{(s)}(t) \,+\, \delta H^{(s)}(t)\right) & \left(\tilde{U}[s|t, t_i] \,+\, \delta\tilde{U}[s|t, t_i]\right) . \end{aligned} \tag{2.15}$$

To find $\delta\tilde{U}$ in terms of $\tilde{U}$ and $\delta H^{(s)}$ we begin by defining

$$R_1 \equiv \tilde{U}, \quad R \equiv \tilde{U} + \delta\tilde{U}, \quad \Lambda_1 \equiv -\frac{i}{\hbar} H^{(s)}, \quad \Lambda \equiv -\frac{i}{\hbar}\delta H^{(s)}, \tag{2.16}$$

which enable us to rewrite Eq.(2.15) in the following compact form

$$R \;=\; I \,+\, (\Lambda + \Lambda_1)\, R \,, \tag{2.17}$$

whereas Eq.(2.11) becomes

$$R_1 \;=\; I \,+\, \Lambda_1 \, R_1 \,. \tag{2.18}$$

By substituting

$$\Lambda_1 \;=\; (R_1 \,-\, I)\, R_1^{-1} \tag{2.19}$$

back into Eq.(2.17) we find

$$R \,-\, R_1 \;=\; R_1 \, \Lambda \, R \,, \tag{2.20}$$

or, in terms of the expanded notation,

$$\delta\tilde{U}[s|t_f, t_i] \;=\; -\frac{i}{\hbar}\int_{t_i}^{t_f} dt \, \tilde{U}[s|t_f, t] \, \delta H^{(s)}(t) \, \tilde{U}[s|t, t_i] \,. \tag{2.21}$$

This is the *Schwinger action principle* [Schwinger (1951c)]. It is the backbone of the *functional differential formulation of Quantum Mechanics* [5].

[5] The derivation of the Schwinger action principle presented here follows along the lines of [Visconti (1965)].

2.3 Green functions

We can verify that

$$\frac{\delta \tilde{U}[s|t_f, t_i]}{\delta J_k(t)} = \frac{i}{\hbar}\,\tilde{U}[s|t_f, t]\,Q^k(t)\,\tilde{U}[s|t, t_i]\,, \quad \forall\; t \in [t_i, t_f]\,, \tag{2.22a}$$

$$\frac{\delta \tilde{U}[s|t_f, t_i]}{\delta K^k(t)} = \frac{i}{\hbar}\,\tilde{U}[s|t_f, t]\,P_k(t)\,\tilde{U}[s|t, t_i]\,, \quad \forall\; t \in [t_i, t_f]\,, \tag{2.22b}$$

are consequences of Eqs.(2.3) and (2.21). Notice that the functional derivatives of the operator $\tilde{U}[s|t_f, t_i]$ with respect to the fictitious sources *do not vanish* at the limit $J = K = 0$. It is also relevant to mention that (for the rules regarding functional differentiation, see appendix B)

$$\frac{\delta J_k(t)}{\delta J_m(t')} = \delta_k^{\ m}\delta(t - t')\,, \tag{2.23a}$$

$$\frac{\delta J_k(t)}{\delta K^m(t')} = 0\,, \tag{2.23b}$$

$$\frac{\delta K^k(t)}{\delta K^m(t')} = \delta^k_{\ m}\delta(t - t')\,, \tag{2.23c}$$

$$\frac{\delta K^k(t)}{\delta J_m(t')} = 0\,, \tag{2.23d}$$

since all fictitious sources are assumed to be linearly independent.

As for the second order functional derivative

$$\frac{\delta^2 \tilde{U}[s|t_f, t_i]}{\delta J_k(t)\delta J_r(t')}\,, \qquad t\,, t' \in [t_i, t_f]\,,$$

we obtain, by starting from Eq.(2.22a),

$$\frac{\delta^2 \tilde{U}[s|t_f, t_i]}{\delta J_k(t)\delta J_r(t')} = \\ \begin{cases} \left(\frac{i}{\hbar}\right)^2\,\tilde{U}[s|t_f, t]\,Q^k(t)\,\tilde{U}[s|t, t']\,Q^r(t')\,\tilde{U}[s|t', t_i] \text{ for } t > t' \\ \left(\frac{i}{\hbar}\right)^2\,\tilde{U}[s|t_f, t']\,Q^r(t')\,\tilde{U}[s|t', t]\,Q^k(t)\,\tilde{U}[s|t, t_i] \text{ for } t' > t \end{cases}. \tag{2.24}$$

A compact writing of the higher order functional derivatives of the operator $\tilde{U}[s|t_f, t_i]$ calls for a new notation. We then introduce

$$\frac{i}{\hbar}\,\Omega^k[s\,|\,t_f, t, t_i] = \frac{i}{\hbar}\,\left(t_f\,,\,Q^k(t)\,, t_i\right) \equiv \frac{\delta \tilde{U}[s|t_f, t_i]}{\delta J_k(t)}\,, \tag{2.25a}$$

$$\frac{i}{\hbar}\,\Omega_k[s\,|\,t_f, t, t_i] = \frac{i}{\hbar}\,\left(t_f\,,\,P_k(t)\,, t_i\right) \equiv \frac{\delta \tilde{U}[s|t_f, t_i]}{\delta K^k(t)}\,, \tag{2.25b}$$

where $\Omega^k[s\,|\,t_f,t,t_i]$ and $\Omega_k[s\,|\,t_f,t,t_i]$ will be referred to as the *one point operator Green functions.* Similarly,

$$\begin{aligned}
&\left(\frac{i}{\hbar}\right)^2 \Omega^{kr}[s\,|\,t_f,t,t',t_i] \\
&= \left(\frac{i}{\hbar}\right)^2 (t_f\,,\,\mathcal{T}\{Q^k(t)\,Q^r(t')\}\,,t_i) \equiv \frac{\delta^2\tilde{U}[s|t_f,t_i]}{\delta J_k(t)\,\delta J_r(t')}\,, \qquad (2.26a) \\
&\left(\frac{i}{\hbar}\right)^2 \Omega_{kr}[s\,|\,t_f,t,t',t_i] \\
&= \left(\frac{i}{\hbar}\right)^2 (t_f\,,\,\mathcal{T}\{P_k(t)\,P_r(t')\}\,,t_i) \equiv \frac{\delta^2\tilde{U}[s|t_f,t_i]}{\delta K^k(t)\,\delta K^r(t')}\,, \qquad (2.26b) \\
&\left(\frac{i}{\hbar}\right)^2 \Omega^k{}_r[s\,|\,t_f,t,t',t_i] \\
&= \left(\frac{i}{\hbar}\right)^2 (t_f\,,\,\mathcal{T}\{Q^k(t)\,P_r(t')\}\,,t_i) \equiv \frac{\delta^2\tilde{U}[s|t_f,t_i]}{\delta J_k(t)\,\delta K^r(t')}\,, \qquad (2.26c)
\end{aligned}$$

designate the *two points operator Green functions.* At last,

$$\begin{aligned}
&\left(\frac{i}{\hbar}\right)^{n+m} \Omega^{k_1,\dots,k_n}_{r_1,\dots,r_m}[s\,|\,t_f,t'_i,\dots,t'_n,t''_1,\dots,t''_m,t_i] \\
&= \left(\frac{i}{\hbar}\right)^{n+m} (t_f\,,\,\mathcal{T}\{Q^{k_1}(t'_i)\cdots Q^{k_n}(t'_n)P_{r_1}(t''_1)\cdots P_{r_m}(t''_m)\}\,,t_i) \\
&\equiv \frac{\delta^{\,n+m}\tilde{U}[s|t_f,t_i]}{\delta J_{k_1}(t'_i)\cdots\delta J_{k_n}(t'_n)\,\delta K^{r_1}(t''_1)\cdots\delta K^{r_m}(t''_m)}\,, \qquad (2.27)
\end{aligned}$$

is the $n+m$ *points operator Green function.* Here, $t'_i,\dots,t'_n$ as well as $t''_1,\dots,t''_m$ belong to the interval $[t_i,t_f]$, while $\mathcal{T}$ denotes the chronological ordering operator.

To proceed we need the results concerning the eigenvalue problem

$$H|E\rangle = E|E\rangle\,. \qquad (2.28)$$

From Eq.(2.10) and for $L = H$ we get

$$\frac{dH(t)}{dt} = 0\,, \qquad (2.29)$$

which asserts that H is time independent. Hence, the eigenvalues in Eq.(2.28) do not depend on time. As for the eigenvectors, they *do not depend upon the fictitious sources* and can be determined up to an arbitrary time dependent phase factor. To find such factor we have to look for the canonical transformation linking the fictitious source picture with the Schrödinger picture. It yields [6]

$$H_S(t) = U_S[s\,|\,t,t_i]\,\tilde{U}^\dagger[s\,|\,t,t_i]\,H\,\tilde{U}[s\,|\,t,t_i]\,U_S^\dagger[s\,|\,t,t_i]\,, \quad (2.30a)$$

$$|E\rangle_S = U_S[s\,|\,t,t_i]\,\tilde{U}^\dagger[s\,|\,t,t_i]\,|E\rangle\,. \quad (2.30b)$$

From Eqs.(2.30) and (2.29) we obtain

$$\frac{dH_S(t)}{dt} = -\frac{i}{\hbar}\left[H_S(t) + H_S^{(s)}(t)\,,\, H_S(t)\right] - \frac{i}{\hbar}\left[H_S(t)\,,\, H_S^{(s)}(t)\right] = 0 \quad (2.31)$$

and

$$H_S\,|E\rangle_S = E\,|E\rangle_S\,. \quad (2.32)$$

Hence, the operator H_S is also time independent and so are the eigenvalues in Eq.(2.32). This confirms that eigenvalues are picture independent.

We fix the eigenvectors of H_S so as to be constant in time. Then, the phase factor of the eigenvectors of H can be found by starting from Eq.(2.30b), i.e.,

$$|E\rangle = \tilde{U}[s\,|\,t,t_i]\,U_S^\dagger[s\,|\,t,t_i]\,|E\rangle_S\,, \quad (2.33)$$

which leads to

$$\frac{d|E\rangle}{dt} = \frac{i}{\hbar}\,E\,|E\rangle \Longrightarrow |E\rangle = |E,t\rangle = e^{\frac{i}{\hbar}E(t-t_i)}\,|E\rangle_S\,. \quad (2.34)$$

To summarize, the eigenvectors of H and H_S are in one-to-one correspondence. Moreover, they are related through a time dependent phase factor depending upon the corresponding eigenvalue.

We are now ready to introduce the $n+m$ *points Green function.* It is defined by

$$\begin{aligned} &\omega^{k_1,\ldots,k_n}_{r_1,\ldots,r_m}\,(t_f, t_i', \ldots, t_n', t_1'', \ldots, t_m'', t_i) \\ &\equiv \langle E_0, t_f)|\,\big(t_f\,, \mathcal{T}\{Q^{k_1}(t_i')\cdots Q^{k_n}(t_n') \\ &\times P_{r_1}(t_1'')\cdots P_{r_m}(t_m'')\}\,, t_i)\,|E_0, t_i)\rangle\big|_{J=K=0} \\ &= \left(\frac{\hbar}{i}\right)^{n+m} \left.\frac{\delta^{\,n+m}\tilde{\mathcal{U}}_0\,[s|t_f, t_i]}{\delta J_{k_1}(t_1')\cdots\delta J_{k_n}(t_n')\delta K^{r_1}(t_1'')\cdots\delta K^{r_m}(t_m'')}\right|_{s=0}, \end{aligned} \quad (2.35)$$

[6] This canonical transformation can be found through a two-step procedure: i) recall Eqs.(1.18) relating the Schrödinger and Heisenberg pictures and ii) use Eqs.(2.6) to go from the Heisenberg to the fictitious source picture.

where

$$\tilde{\mathcal{U}}_0\left[s|t_f,t_i\right] \equiv \langle E_0,t_f|\tilde{U}[s|t_f,t_i]|E_0,t_i\rangle \tag{2.36}$$

is the *Green functions generating functional.* Here, $|E_0,t\rangle$ is the ground state of H. As we already pointed out, it does not depend on the fictitious sources and, therefore, goes freely through the operations of functional differentiation. Also, we assume H to be bounded from below which implies that $|E_0| < \infty$. The physical meaning of $\tilde{\mathcal{U}}_0\left[s|t_f,t_i\right]$ is clear: it is the *ground state persistence amplitude when the physical system is acted upon by fictitious sources.*

Next we introduce the $n+m$ points *normalized Green functions* as

$$\begin{aligned}&\omega^{k_1,\ldots,k_n}_{(N)\ r_1,\ldots,r_m}\left(t_f,t'_i,\ldots,t'_n,t''_1,\ldots,t''_m,t_i\right)\\ &\equiv \frac{1}{\tilde{\mathcal{U}}_0\left[s=0|t_f,t_i\right]}\,\omega^{k_1,\ldots,k_n}_{\ r_1,\ldots,r_m}\left(t_f,t'_i,\ldots,t'_n,t''_1,\ldots,t''_m,t_i\right).\end{aligned} \tag{2.37}$$

From Eqs.(2.12) and (2.36) it follows that

$$\begin{aligned}\tilde{\mathcal{U}}_0\left[s=0|t_f,t_i\right] &= \langle E_0,t_f|\tilde{U}[s=0|t_f,t_i]|E_0,t_i\rangle\\ = \langle E_0,t_f|E_0,t_i\rangle &= e^{-\frac{i}{\hbar}E_0(t_f-t_i)}.\end{aligned} \tag{2.38}$$

However, we are allowed to set

$$E_0 = 0\,, \tag{2.39}$$

since it only requires an overall shifting of the energy spectrum. Hence, we shall next adopt the normalization condition

$$\tilde{\mathcal{U}}_0\left[s=0|t_f,t_i\right] = 1\,. \tag{2.40}$$

With this choice, normalized and ordinary Green functions become identical.

However, physical amplitudes are associated with the *connected Green functions*. By definition, the $n+m$ points *connected Green function* is given by

$$\begin{aligned}&\omega^{k_1\ldots k_n}_{(C)\ r_1\ldots r_m}\left(t_f,t'_i,\ldots,t'_n,t''_1,\ldots,t''_m,t_i\right)\\ &\equiv \left(\frac{\hbar}{i}\right)^{n+m-1}\left.\frac{\delta^{\,n+m}\tilde{W}_0\left[s|t_f,t_i\right]}{\delta J_{k_1}(t'_1)\cdots\delta J_{k_n}(t'_n)\delta K^{r_1}(t''_1)\cdots\delta K^{r_m}(t''_m)}\right|_{s=0},\end{aligned} \tag{2.41}$$

where $\tilde{W}_0[s|t_f, t_i]$ is the *connected Green functions generating functional.* It is defined by

$$\tilde{\mathcal{U}}_0\,[s|t_f, t_i] \;\equiv:\; e^{\frac{i}{\hbar}\tilde{W}_0[s|t_f,t_i]} \tag{2.42}$$

or, equivalently,

$$\tilde{W}_0[s|t_f, t_i] \;\equiv\; \frac{\hbar}{i}\,\ln\tilde{\mathcal{U}}_0\,[s|t_f, t_i]\,. \tag{2.43}$$

It is instructive to express the connected Green function in terms of the normalized ones. For the one and two point connected Green functions we find, respectively,

$$\omega^{k_1}_{(C)}(t_f, t'_1, t_i) \;=\; \left.\frac{\delta\tilde{W}_0[s|t_f, t_i]}{\delta J_{k_1}(t'_1)}\right|_{s=0} \;=\; \frac{\hbar}{i}\,\frac{1}{\tilde{\mathcal{U}}_0\,[s|t_f, t_i]}\,\left.\frac{\delta\tilde{\mathcal{U}}_0\,[s|t_f, t_i]}{\delta J_{k_1}(t'_1)}\right|_{s=0}$$
$$= \omega^{k_1}(t_f, t'_1, t_i) \tag{2.44}$$

and

$$\omega^{k_1}_{(C)\;r_1}(t_f, t'_1, t''_1, t_i) \;=\; \frac{\hbar}{i}\,\left.\frac{\delta^2\tilde{W}_0[s|t_f, t_i]}{\delta J_{k_1}(t'_1)\delta K^{r_1}(t''_1)}\right|_{s=0}$$
$$= \left(\frac{\hbar}{i}\right)^2\left[\frac{1}{\tilde{\mathcal{U}}_0\,[s|t_f, t_i]}\,\left.\frac{\delta^2\tilde{\mathcal{U}}_0\,[s|t_f, t_i]}{\delta J_{k_1}(t'_1)\delta K^{r_1}(t''_1)}\right|_{s=0}\right.$$
$$\left.-\frac{1}{\tilde{\mathcal{U}}_0^2\,[s|t_f, t_i]}\,\frac{\delta\tilde{\mathcal{U}}_0\,[s|t_f, t_i]}{\delta J_{k_1}(t'_1)}\,\left.\frac{\delta\tilde{\mathcal{U}}_0\,[s|t_f, t_i]}{\delta K^{r_1}(t''_1)}\right|_{s=0}\right]$$
$$= \omega^{k_1}_{r_1}\left(t_f, t'_i, t''_1, t_i\right) - \omega^{k_1}(t_f, t'_1, t_i)\,\omega_{r_1}(t_f, t''_1, t_i)\,. \tag{2.45}$$

The definition (2.42) eliminates the disconnected parts from

$$\omega^{k_1\ldots k_n}_{r_1\ldots r_m}\left(t_f, t'_i, \ldots, t'_n, t''_1, \ldots, t''_m, t_i\right)\,.$$

It can be observed at work in Eq.(2.45).

Solving the quantum dynamics of a physical system implies in finding the corresponding infinite set of connected Green functions. Since all Green functions arise from differentiating the corresponding generating functional with respect to the fictitious sources, we shall focus on solving the functional differential equations obeyed by $\tilde{\mathcal{U}}_0\,[s|t_f, t_i]$, namely, *Schwinger equations* [Schwinger (1951a,b)].

2.4 Schwinger equations

The functional dependence of $\tilde{U}[s|t_f, t_i]$ upon the fictitious sources implies that we shall be dealing with functional differential equations. This may appear to be complicated at first sight. However, one is to recall that $H^{(s)}$ depends linearly upon J and K (see Eq.(2.3)). This lowers the level of complexity since only *functionally linear* differential equations will be involved. In short, the dependence of H on Q and P is nonlinear while that of $H + H^{(s)}$ on J and K is linear. This is the landmark of the formalism put forward by Schwinger.

We start by considering the one point operator Green function in Eq.(2.22a) and focus on computing

$$\frac{d}{dt}\left\{\frac{\delta\tilde{U}[s|t_f,t_i]}{\delta J_l(t)}\right\} = \frac{d}{dt}\left\{\frac{i}{\hbar}\tilde{U}[s|t_f,t]\,Q^l(t)\,\tilde{U}[s|t,t_i]\right\}. \tag{2.46}$$

All three terms in the right hand side contribute to d/dt. The calculation is facilitated by writing Q^l in terms of its Heisenberg picture image Q^l_H. On account of Eq.(2.6a) we can write

$$\begin{aligned}\frac{d}{dt}\left\{\frac{\delta\tilde{U}[s|t_f,t_i]}{\delta J_l(t)}\right\} &= \frac{d}{dt}\left\{\frac{i}{\hbar}\tilde{U}[s|t_f,t_0]\,Q^l_H(t)\,\tilde{U}[s|t_0,t_i]\right\} \\ &= \frac{i}{\hbar}\,\tilde{U}[s|t_f,t_0]\,\frac{dQ^l_H(t)}{dt}\,\tilde{U}[s|t_0,t_i]\,,\end{aligned} \tag{2.47}$$

since the t-dependence is now fully concentrated on $Q^l_H(t)$. By invoking the Heisenberg equation of motion (2.4) we arrive at

$$\begin{aligned}&\frac{d}{dt}\left\{\frac{\hbar}{i}\frac{\delta\tilde{U}[s|t_f,t_i]}{\delta J_l(t)}\right\} \\ &= \frac{i}{\hbar}\,\tilde{U}[s|t_f,t_0]\,\left[H_H(t)\,,\,Q^l_H(t)\right]\,\tilde{U}[s|t_0,t_i] \\ &+ \frac{i}{\hbar}\,\tilde{U}[s|t_f,t_0]\,\left[H^{(s)}_H(t)\,,\,Q^l_H(t)\right]\,\tilde{U}[s|t_0,t_i]\,,\end{aligned} \tag{2.48}$$

which, by returning to the fictitious source picture, goes into

$$\begin{aligned}&\frac{d}{dt}\left\{\frac{\hbar}{i}\frac{\delta\tilde{U}[s|t_f,t_i]}{\delta J_l(t)}\right\} \\ &= \frac{i}{\hbar}\,\tilde{U}[s|t_f,t]\,\left[H\,,\,Q^l(t)\right]\,\tilde{U}[s|t,t_i] \\ &+ \frac{i}{\hbar}\,\tilde{U}[s|t_f,t]\,\left[H^{(s)}(t)\,,\,Q^l(t)\right]\,\tilde{U}[s|t,t_i]\,.\end{aligned} \tag{2.49}$$

Correspondingly, we obtain

$$\begin{aligned}
&\frac{d}{dt}\left\{\frac{\hbar}{i}\frac{\delta\tilde{U}[s|t_f,t_i]}{\delta K^l(t)}\right\} \\
&= \frac{i}{\hbar}\tilde{U}[s|t_f,t]\,[H\,,\,P_l(t)]\,\tilde{U}[s|t,t_i] \\
&+ \frac{i}{\hbar}\tilde{U}[s|t_f,t]\,\left[H^{(s)}(t)\,,\,P_l(t)\right]\,\tilde{U}[s|t,t_i]\,. \qquad (2.50)
\end{aligned}$$

The structure of $H^{(s)}$ (see Eq.(2.3)) as well as the equal time algebra obeyed by the basic observables (see Eq.(2.1)) are known. When this information is brought into Eqs.(2.49) and (2.50) they become, respectively,

$$\begin{aligned}
&\frac{d}{dt}\left\{\frac{\hbar}{i}\frac{\delta\tilde{U}[s|t_f,t_i]}{\delta J_l(t)}\right\} \\
&= \tilde{U}[s|t_f,t]\,A^l\left(Q(t),P(t)\right)\,\tilde{U}[s|t,t_i]\;-\;K^l(t)\,\tilde{U}[s|t_f,t_i]\,, \qquad (2.51a) \\
&\frac{d}{dt}\left\{\frac{\hbar}{i}\frac{\delta\tilde{U}[s|t_f,t_i]}{\delta K^l(t)}\right\} \\
&= \tilde{U}[s|t_f,t]\,B_l\left(Q(t),P(t)\right)\,\tilde{U}[s|t,t_i]\;+\;J_l(t)\,\tilde{U}[s|t_f,t_i]\,, \qquad (2.51b)
\end{aligned}$$

where

$$A^l\left(Q(t),P(t)\right) \equiv \frac{i}{\hbar}\,\left[H\,,\,Q^l(t)\right] \;=\; \frac{\partial H}{\partial P_l}\,, \qquad (2.52a)$$

$$B_l\left(Q(t),P(t)\right) \equiv \frac{i}{\hbar}\,\left[H\,,\,P_l(t)\right] \;=\; -\frac{\partial H}{\partial Q^l}\,. \qquad (2.52b)$$

In order to proceed we must specify the structure of the composite operator $H = H(Q,P)$. When $A^l\left(Q(t),P(t)\right)$ and $B_l\left(Q(t),P(t)\right)$ can unambiguously be written in terms of the functional derivatives of $\tilde{U}[s|t_f,t_i]$ with respect to the fictitious sources, the setting of Schwinger equations will be over. Meanwhile, we shall restrict ourselves to deal with systems whose Hamiltonian is of the form

$$H(Q,P) \;=\; F(Q) + G(P)\,. \qquad (2.53)$$

Correspondingly, Eqs.(2.51) get down to

$$\frac{d}{dt}\left\{\frac{\hbar}{i}\frac{\delta\tilde{U}[s|t_f,t_i]}{\delta J_l(t)}\right\}$$
$$= \tilde{U}[s|t_f,t]\,A^l(P(t))\,\tilde{U}[s|t,t_i] \,-\, K^l(t)\,\tilde{U}[s|t_f,t_i]\,, \tag{2.54a}$$
$$\frac{d}{dt}\left\{\frac{\hbar}{i}\frac{\delta\tilde{U}[s|t_f,t_i]}{\delta K^l(t)}\right\}$$
$$= \tilde{U}[s|t_f,t]\,B_l(Q(t))\,\tilde{U}[s|t,t_i] \,+\, J_l(t)\,\tilde{U}[s|t_f,t_i]\,, \tag{2.54b}$$

with

$$A^l\left(P(t)\right) = \frac{\partial G(P)}{\partial P_l}\,, \tag{2.55a}$$
$$B_l\left(P(t)\right) = -\frac{\partial F(Q)}{\partial Q^l}\,. \tag{2.55b}$$

Thus, the question now is whether or not

$$\tilde{U}[s|t_f,t]\,A^l(P(t))\,\tilde{U}[s|t,t_i] \quad\text{and}\quad \tilde{U}[s|t_f,t]\,B_l(Q(t))\,\tilde{U}[s|t,t_i] \tag{2.56}$$

can be written as the functional derivatives of $\tilde{U}[s|t_f,t_i]$ with respect to the fictitious sources. We shall begin by considering the case raised by the generic operator function $f(Q)$, where the observables $Q^1,\ldots,Q^N$ are Cartesian coordinates with the following common set of eigenvectors

$$\{|q^1,\ldots,q^l,\ldots,q^N\rangle;\quad -\infty\le q^l\le+\infty,\quad l=1,\ldots,N\}\,. \tag{2.57}$$

The spectral resolution of $f(Q)$ then reads

$$f(Q) = \left(\prod_{l=1}^{N}\int_{-\infty}^{+\infty}dq^l\right)|q^1,\ldots,q^N\rangle\,f(q^1,\ldots,q^N)\,\langle q^1,\ldots,q^N|\,. \tag{2.58}$$

By assumption, the function $f(q^1,\ldots,q^N)$ is analytic in a neighborhood of the origin. Hence, $f(q)$ may be replaced by its Maclaurin expansion within the region of analyticity. The n-th term in this expansion is a set of polynomials of the form $q^{l_1}\cdots q^{l_n}$. Thus, the problem reduces to find $\tilde{U}[s|t_f,t]\,Q^{l_1}(t)\cdots Q^{l_n}(t)\tilde{U}[s|t,t_i]$ in terms of the functional derivatives of $\tilde{U}[s|t_f,t_i]$ with respect to the fictitious sources. For the polynomials of degree *zero* and *one* this is certainly possible since

$$\tilde{U}[s|t_f,t]\,\tilde{U}[s|t,t_i] = \tilde{U}[s|t_f,t_i] \tag{2.59}$$

and

$$\tilde{U}[s|t_f,t]\,Q^l(t)\,\tilde{U}[s|t,t_i] \;=\; \big(t_f,Q^l(t),t_i\big) \;=\; \frac{\hbar}{i}\,\frac{\delta\tilde{U}[s|t_f,t_i]}{\delta J_l(t)}\,, \tag{2.60}$$

where Eq.(2.25a) was taken into account. For the second and higher degree polynomials the situation is more involved. In fact, the object that can be written in terms of the functional derivatives of $\tilde{U}[s|t_f,t_i]$ with respect to the fictitious sources is

$$(t_f,\mathcal{T}\{Q^{l_1}(t_1)\dots Q^{l_n}(t_n)\},t_i)\,.$$

In particular

$$\begin{aligned}
&\left(\frac{\hbar}{i}\right)^2 \frac{\delta^2\tilde{U}[s|t_f,t_i]}{\delta J_l(t)\delta J_m(t')} \;=\; \big(t_f,\mathcal{T}\{Q^l(t)Q^m(t')\},t_i\big)\\
&= \theta(t-t')\,\tilde{U}[s|t_f,t]\,Q^l(t)\,\tilde{U}[s|t,t']\,Q^m(t')\,\tilde{U}[s|t',t_i]\\
&+\theta(t'-t)\,\tilde{U}[s|t_f,t']\,Q^m(t')\,\tilde{U}[s|t',t]\,Q^l(t))\,\tilde{U}[s|t,t_i]\,,
\end{aligned} \tag{2.61}$$

where $\theta(t)$ denotes the generalized Heaviside function. Furthermore,

$$\lim_{t'\uparrow t}\big(t_f,\mathcal{T}\{Q^l(t)Q^m(t')\},t_i\big) \;=\; \tilde{U}[s|t_f,t]\,Q^l(t)\,Q^m(t)\,\tilde{U}[s|t,t_i]\,, \tag{2.62a}$$

$$\lim_{t'\downarrow t}\big(t_f,\mathcal{T}\{Q^l(t)Q^m(t')\},t_i\big) \;=\; \tilde{U}[s|t_f,t]\,Q^m(t)\,Q^l(t)\,\tilde{U}[s|t,t_i]\,, \tag{2.62b}$$

which are, in principle, different operators. However, the commutation rules in Eq.(2.1a) state that the equal time commutator involving two position observables vanishes securing that $\big(t_f,\mathcal{T}\{Q^l(t)Q^m(t')\},t_i\big)$ is continuous at $t = t'$. This is just one of the second degree polynomials entering the Maclaurin expansion above. Hence, we are entitled to write

$$\tilde{U}[s|t_f,t]\,Q^l(t)\,Q^m(t)\,\tilde{U}[s|t,t_i] \;=\; \left(\frac{\hbar}{i}\right)^2 \frac{\delta^2\tilde{U}[s|t_f,t_i]}{\delta J_l(t)\delta J_m(t)}\,. \tag{2.63}$$

One can convince oneself that these arguments also apply for the higher degree polynomials and, therefore,

$$\tilde{U}[s|t_f,t]\,f(Q(t))\,\tilde{U}[s|t,t_i] \;=\; f\left[\frac{\hbar}{i}\frac{\delta}{\delta J(t)}\right]\,\tilde{U}[s|t_f,t_i]\,. \tag{2.64}$$

Similarly,

$$\tilde{U}[s|t_f, t]\, g(P(t))\, \tilde{U}[s|t, t_i] = g\left[\frac{\hbar}{i}\frac{\delta}{\delta K(t)}\right] \tilde{U}[s|t_f, t_i]\,. \tag{2.65}$$

Then, Eqs.(2.54a) and (2.54b) can be rewritten, respectively, as

$$\frac{d}{dt}\left\{\frac{\hbar}{i}\frac{\delta\tilde{U}[s|t_f, t_i]}{\delta J_l(t)}\right\}$$

$$= A^l\left(\frac{\hbar}{i}\frac{\delta}{\delta K^l(t)}\right)\tilde{U}[s|t_f, t_i] - K^l(t)\,\tilde{U}[s|t_f, t_i]\,, \tag{2.66a}$$

$$\frac{d}{dt}\left\{\frac{\hbar}{i}\frac{\delta\tilde{U}[s|t_f, t_i]}{\delta K^l(t)}\right\}$$

$$= B_l\left(\frac{\hbar}{i}\frac{\delta}{\delta J_l(t)}\right)\tilde{U}[s|t_f, t_i] + J_l(t)\,\tilde{U}[s|t_f, t_i]\,. \tag{2.66b}$$

By taking matrix elements on both sides of Eqs.(2.66) between the states $|E_0, t_i\rangle$ and $|E_0, t_f\rangle$ and by using, afterwards, the definition in Eq.(2.36) we arrive at

$$\frac{d}{dt}\left\{\frac{\hbar}{i}\frac{\delta\tilde{\mathcal{U}}_0\,[s|t_f, t_i]}{\delta J_l(t)}\right\}$$

$$= A^l\left(\frac{\hbar}{i}\frac{\delta}{\delta K^l(t)}\right)\tilde{\mathcal{U}}_0\,[s|t_f, t_i] - K^l(t)\tilde{\mathcal{U}}_0\,[s|t_f, t_i]\,, \tag{2.67a}$$

$$\frac{d}{dt}\left\{\frac{\hbar}{i}\frac{\delta\tilde{\mathcal{U}}_0\,[s|t_f, t_i]}{\delta K^l(t)}\right\}$$

$$= B_l\left(\frac{\hbar}{i}\frac{\delta}{\delta J_l(t)}\right)\tilde{\mathcal{U}}_0\,[s|t_f, t_i] + J_l(t)\tilde{\mathcal{U}}_0\,[s|t_f, t_i]\,, \tag{2.67b}$$

which are *Schwinger equations* [Schwinger (1951a,b)] [7].

2.5 Integration of Schwinger equations

The fictitious sources linearly enter the Schwinger equations (see Eqs.(2.67)). It is therefore natural to attempt their integration by making use of the functional Fourier transform (see appendix B). To implement

[7] It is worth mentioning the existence of an alternative approach for deriving Schwinger equations due to Symanzik [Symanzik (1954)].

this approach, we shall first try to grasp about the structure of the solution by exploring the relationship linking $\tilde{\mathcal{U}}_0[s|t_f,t_i]$ with ${}_H\langle q_f,t_f|q_i,t_i\rangle_H^{(s)}$, namely,

$$\begin{aligned}\tilde{\mathcal{U}}_0\,[s|t_f,t_i] &= \langle E_0,t_f|\tilde{U}[s|t_f,t_i]|E_0,t_i\rangle = \int_{-\infty}^{+\infty} d^N q_f \int_{-\infty}^{+\infty} d^N q_i \\ &= \langle E_0,t_f|q_f,t_f\rangle\langle q_f,t_f|\tilde{U}[s|t_f,t_i]|q_i,t_i\rangle\langle q_i,t_i|E_0,t_i\rangle \\ &= \int_{-\infty}^{+\infty} d^N q_f \int_{-\infty}^{+\infty} d^N q_i\, \psi_0^*(q_f)\,{}_H\langle q_f,t_f|q_i,t_i\rangle_H^{(s)}\,\psi_0(q_i)\,. \end{aligned} \tag{2.68}$$

In the first line of this equality the definition in Eq.(2.36) was used; the last line, however, is a consequence of Eq.(2.6b). Indeed,

$${}_H\langle q_f,t_f|q_i,t_i\rangle_H^{(s)} = \langle q_f,t_f|\tilde{U}[s|t_f,t_i]|q_i,t_i\rangle\,. \tag{2.69}$$

Furthermore, $\psi_0(q)$ denotes the ground state energy eigenfunction. In particular,

$$\psi_0^*(q_f) = \langle E_0,t_f|q_f,t_f\rangle = {}_S\langle E_0|q_f\rangle_S\,, \tag{2.70a}$$

$$\psi_0(q_i) = \langle q_i,t_i|E_0,t_i\rangle = {}_S\langle q_i|E_0\rangle_S\,. \tag{2.70b}$$

We shall, therefore, be seeking for a solution of the form

$$\begin{aligned}\tilde{\mathcal{U}}_0\,[s|t_f,t_i] &= \int_{-\infty}^{+\infty} d^N q_f \int_{-\infty}^{+\infty} d^N q_i\, \psi_0^*(q_f)\,\psi_0(q_i) \int [\mathcal{D}q] \int [\mathcal{D}p] \\ &\times \tilde{\mathcal{U}}_0[\tilde{s}|t_f,t_i] \exp\left\{\frac{i}{\hbar}\int_{t_i}^{t_f} d\tau \sum_{j=1}^{N}\left[q^j(\tau)J_j(\tau) + p_j(\tau)K^j(\tau)\right]\right\}\,, \end{aligned} \tag{2.71}$$

where $\tilde{\mathcal{U}}_0\,[\tilde{s}|t_f,t_i]$ denotes the functional Fourier transform of $\tilde{\mathcal{U}}_0\,[s|t_f,t_i]$ whereas

$$\tilde{s} \equiv q^j\,, p_j\,, \qquad j = 1,\ldots,N\,, \tag{2.72}$$

are the variables conjugate to the fictitious sources $s \equiv J, K$. In this book, q and p designate coordinates and momenta, respectively. However, they are presently being used for labeling the variables assembled in $\tilde{s}$. This is not an *abuse* of notation because it will shortly be shown that the variables conjugate to J and K are, in fact, q and p, respectively. As for the propagator in the presence of the fictitious sources, the solution in Eq.(2.71) implies that

$$
\begin{aligned}
{}_H\langle q_f, t_f | q_i, t_i \rangle_H^{(s)} &= \int [\mathcal{D}q] \int [\mathcal{D}p]\, \tilde{\mathcal{U}}_0 \left[\tilde{s} | t_f, t_i\right] \\
&\times \exp\left\{ \frac{i}{\hbar} \int_{t_i}^{t_f} d\tau \sum_{k=1}^{N} \left[q^j(\tau) J_j(\tau) + p_j(\tau) K^j(\tau) \right] \right\} . \qquad (2.73)
\end{aligned}
$$

We now return with Eq.(2.71) into Eqs.(2.67). As far as Eq.(2.67a) is concerned, after some algebra, the following results arise:

$$
\begin{aligned}
&\frac{d}{dt} \left\{ \frac{\hbar}{i} \frac{\delta \tilde{\mathcal{U}}_0 \left[s | t_f, t_i\right]}{\delta J_l(t)} \right\} = \int_{-\infty}^{+\infty} d^N q_f \int_{-\infty}^{+\infty} d^N q_i\, \psi_0^*(q_f)\, \psi_0(q_i) \int [\mathcal{D}q] \int [\mathcal{D}p] \\
&\times \tilde{\mathcal{U}}_0[\tilde{s}|t_f, t_i] \dot{q}^l(t) \exp\left\{ \frac{i}{\hbar} \int_{t_i}^{t_f} d\tau \sum_{j=1}^{N} \left[q^j(\tau) J_j(\tau) + p_j(\tau) K^j(\tau) \right] \right\} , \qquad (2.74)
\end{aligned}
$$

$$
\begin{aligned}
&A^l \left(\frac{\hbar}{i} \frac{\delta}{\delta K^l(t)} \right) \tilde{\mathcal{U}}_0 \left[s | t_f, t_i\right] = \int_{-\infty}^{+\infty} d^N q_f \int_{-\infty}^{+\infty} d^N q_i\, \psi_0^*(q_f)\, \psi_0(q_i) \int [\mathcal{D}q] \int [\mathcal{D}p] \\
&\times \tilde{\mathcal{U}}_0[\tilde{s}|t_f, t_i]\, A^l(p) \exp\left\{ \frac{i}{\hbar} \int_{t_i}^{t_f} \sum_{j=1}^{N} \left[q^j(\tau) J_j(\tau) + p_j(\tau) K^j(\tau) \right] \right\} \\
&= \int_{-\infty}^{+\infty} d^N q_f \int_{-\infty}^{+\infty} d^N q_i\, \psi_0^*(q_f)\, \psi_0(q_i) \int [\mathcal{D}q] \int [\mathcal{D}p] \tilde{\mathcal{U}}_0[\tilde{s}|t_f, t_i] \\
&\times \frac{\partial h(q(t), p(t))}{\partial p_l(t)} \exp\left\{ \frac{i}{\hbar} \int_{t_i}^{t_f} d\tau \sum_{j=1}^{N} \left[q^j(\tau) J_j(\tau) + p_j(\tau) K^j(\tau) \right] \right\} , \qquad (2.75)
\end{aligned}
$$

where Eq.(2.52a) was taken into account. Furthermore,

$$
\begin{aligned}
-K^l(t) \tilde{\mathcal{U}}_0 \left[s | t_f, t_i\right] &= - \int_{-\infty}^{+\infty} d^N q_f \int_{-\infty}^{+\infty} d^N q_i\, \psi_0^*(q_f)\, \psi_0(q_i) \\
&\times \int [\mathcal{D}q] \int [\mathcal{D}p] \tilde{\mathcal{U}}_0[\tilde{s}|t_f, t_i] \frac{\hbar}{i} \frac{\delta}{\delta p_l(t)} \\
&\times \left(\exp\left\{ \frac{i}{\hbar} \int_{t_i}^{t_f} d\tau \sum_{j=1}^{N} \left[q^j(\tau) J_j(\tau) + p_j(\tau) K^j(\tau) \right] \right\} \right) , \qquad (2.76)
\end{aligned}
$$

which after a by part integration becomes

$$- K^l(t)\tilde{\mathcal{U}}_0\,[s|t_f,t_i] \;=\; + \int_{-\infty}^{+\infty} d^N q_f \int_{-\infty}^{+\infty} d^N q_i\, \psi_0^*(q_f)\,\psi_0(q_i) \int [\mathcal{D}q] \int [\mathcal{D}p]$$
$$\times \left\{ \frac{\delta}{\delta p_l(t)} \tilde{\mathcal{U}}_0[\tilde{s}|t_f,t_i] \right\}$$
$$\times \frac{\hbar}{i} \exp\left\{ \frac{i}{\hbar} \int_{t_i}^{t_f} d\tau \sum_{j=1}^{N} \left[q^j(\tau)J_j(\tau) + p_j(\tau)K^j(\tau) \right] \right\} . \tag{2.77}$$

The same systematics applies for Eq.(2.67b). By collecting these results we arrive at the following homogeneous system of functional differential equations

$$\frac{i}{\hbar}\left[\dot{q}^l(t) \;-\; \frac{\partial h(q(t),p(t))}{\partial p_l(t)} \right] \tilde{\mathcal{U}}_0\,[\tilde{s}|t_f,t_i] \;=\; \frac{\delta \tilde{\mathcal{U}}_0\,[\tilde{s}|t_f,t_i]}{\delta p_l(t)}\,, \tag{2.78a}$$
$$\frac{i}{\hbar}\left[-\dot{p}_l(t) \;-\; \frac{\partial h(q(t),p(t))}{\partial q^l(t)} \right] \tilde{\mathcal{U}}_0\,[\tilde{s}|t_f,t_i] \;=\; \frac{\delta \tilde{\mathcal{U}}_0\,[\tilde{s}|t_f,t_i]}{\delta q^l(t)}\,, \tag{2.78b}$$

whose solution, up to an arbitrary constant $\mathcal{N}$, is

$$\tilde{\mathcal{U}}_0[\tilde{s}|t_f,t_i] = \mathcal{N} \exp\left\{ \frac{i}{\hbar} \int_{t_i}^{t_f} dt \left[\sum_{j=1}^{N} p_j(t)\dot{q}^j(t) - h(q(t),p(t)) \right] \right\} . \tag{2.79}$$

Moreover, in the expressions above $h(q,p)$ refers to the c-number function deriving from the Hamiltonian operator through the mapping

$$H(Q(t),P(t))\Big|_{Q\to q\,,P\to p} \longrightarrow h(q(t),p(t))\,. \tag{2.80}$$

Hence, $h(q,p)$ is the classical Hamiltonian. This justifies the notation in Eq.(2.72).

By returning with Eq.(2.79) into Eqs.(2.71) and (2.73) we obtain, respectively,

$$\tilde{\mathcal{U}}_0\,[s|t_f,t_i] \;=\; \mathcal{N} \int_{-\infty}^{+\infty} d^N q_f \int_{-\infty}^{+\infty} d^N q_i\, \psi_0^*(q_f)\,\psi_0(q_i) \int [\mathcal{D}q] \int [\mathcal{D}p]$$
$$\times \exp\left\{ \frac{i}{\hbar} \int_{t_i}^{t_f} dt \left[\sum_{j=1}^{N} p_j(t)\dot{q}^j(t) \;-\; h(q(t),p(t)) \right] \right.$$
$$\left. + \frac{i}{\hbar} \int_{t_i}^{t_f} dt \sum_{j=1}^{N} \left[q^j(t)J_j(t) \;+\; p_j(t)K^j(t) \right] \right\} \tag{2.81}$$

and

$$
\begin{aligned}
{}_H\langle q_f, t_f | q_i, t_i \rangle_H^{(s)} &= \mathcal{N} \int [\mathcal{D}q] \int [\mathcal{D}p] \\
&\times \exp \left\{ \frac{i}{\hbar} \int_{t_i}^{t_f} dt \left[\sum_{j=1}^{N} p_j(t) \dot{q}^j(t) - h(q(t), p(t)) \right] \right. \\
&\left. + \frac{i}{\hbar} \int_{t_i}^{t_f} dt \sum_{j=1}^{N} \left[q^j(t) J_j(t) + p_j(t) K^j(t) \right] \right\}, \qquad (2.82)
\end{aligned}
$$

where the constant $\mathcal{N}$ is to be determined from Eq.(2.40). This yields

$$
\begin{aligned}
\mathcal{N}^{-1} &= \int_{-\infty}^{+\infty} d^N q_f \int_{-\infty}^{+\infty} d^N q_i \, \psi_0^*(q_f) \, \psi_0(q_i) \int [\mathcal{D}q] \int [\mathcal{D}p] \\
&\times \exp \left\{ \frac{i}{\hbar} \int_{t_i}^{t_f} dt \left[\sum_{j=1}^{N} p_j(t) \dot{q}^j(t) - h(q(t), p(t)) \right] \right\}. \qquad (2.83)
\end{aligned}
$$

What we have in the right hand side of Eq.(2.82) is the phase space path integral. Thus, *the phase space path integral is the functional Fourier transform solving Schwinger equations.* Indeed, both the differential and integral functional formulations of quantum mechanics provide alternative descriptions of the same physical reality.

We take advantage of Eqs.(2.37) and (2.81) to obtain the explicit forms for the Green functions. They read:

$$
\begin{aligned}
&\omega^{k_1, \ldots, k_n}_{r_1, \ldots, r_m} \left(t_f, t_i', \ldots, t_n', t_1'', \ldots, t_m'', t_i \right) \\
&= \mathcal{N} \int_{-\infty}^{+\infty} d^N q_f \int_{-\infty}^{+\infty} d^N q_i \, \psi_0^*(q_f) \, \psi_0(q_i) \int [\mathcal{D}q] \int [\mathcal{D}p] \\
&\times q^{k_1}(t_i') \ldots q^{k_n}(t_n') \, p_{r_1}(t_i'') \ldots p_{r_m}(t_m'') \\
&\times \exp \left\{ \frac{i}{\hbar} \int_{t_i}^{t_f} dt \left[\sum_{j=1}^{N} p_j(t) \dot{q}^j(t) - h(q(t), p(t)) \right] \right\}. \qquad (2.84)
\end{aligned}
$$

2.6 Schwinger equations and the GWT

We now face the problem of setting Schwinger equations in the more general case, i.e., when the structure of the Hamiltonian operator contains products of noncommuting operators. Although the starting point is again Eq.(2.51) we cannot anticipate the possibility of writing $A^l(Q(t), P(t))$ and $B_l(Q(t), P(t))$ as functional derivatives of $\tilde{U}[s|t_f, t_i]$ with respect to the fictitious sources.

A way out from the trouble was proposed in [Girotti and Simões (1980)] and it is based on the property of the GWT quoted in Eq.(1.66c) which, along with Eq.(1.56), leads to

$$\begin{Bmatrix} A^l\,(Q(t), P(t)) \\ B_l\,(Q(t), P(t)) \end{Bmatrix} = \begin{Bmatrix} A^l_{\alpha,-\alpha}\,(Q(t), P(t)) \\ B_{l_{\alpha,-\alpha}}\,(Q(t), P(t)) \end{Bmatrix}$$

$$= \int_{-\infty}^{+\infty} d^N x d^N y \begin{Bmatrix} a^l_\alpha(x,y) \\ b_{l_\alpha}(x,y) \end{Bmatrix} \Delta_{-\alpha}(Q - x, P - y)\,. \tag{2.85}$$

Here, the c-functions a^l_α and b_{l_α} are the GWT of the operators A^l and B_l, respectively. Furthermore, from Eq.(1.51) we find

$$\Delta_{-\alpha}(Q - x, P - y) = (2\pi\hbar)^{-2N}$$

$$\times \int_{-\infty}^{+\infty} d^N u\, d^N v\, e^{\frac{i}{\hbar}(\frac{1}{2}-\alpha)u\cdot v}\, e^{\frac{i}{\hbar}(x-Q(t))\cdot u} e^{\frac{i}{\hbar}(y-P(t))\cdot v}\,. \tag{2.86}$$

What shows up in the right hand side of this last equation is the product of a function of Q times a function of P. This is not a chronological product, though. However,

$$e^{\frac{i}{\hbar}(x-Q(t))\cdot u} e^{\frac{i}{\hbar}(y-P(t))\cdot v} = \lim_{t'\uparrow t} \mathcal{T}\left\{ e^{\frac{i}{\hbar}(x-Q(t))\cdot u} e^{\frac{i}{\hbar}\left(y-P(t')\right)\cdot v} \right\}\,. \tag{2.87}$$

By returning with Eq.(2.87) into Eq.(2.86) and, afterwards, into Eqs.(2.85) and (2.51) we arrive at

$$\frac{d}{dt}\left\{ \frac{\hbar}{i} \frac{\delta \tilde{\mathcal{U}}_0\,[s|t_f, t_i]}{\delta J_l(t)} \right\}$$

$$= \int_{-\infty}^{+\infty} d^N x\, d^N y\, a^l_\alpha(x,y)\, \Delta_{-\alpha}\left(\frac{\hbar}{i}\frac{\delta}{\delta J(t)} - x\,,\, \frac{\hbar}{i}\frac{\delta}{\delta K(t)} - y \right) \tilde{\mathcal{U}}_0\,[s|t_f, t_i]$$

$$- K^l(t)\, \tilde{\mathcal{U}}_0\,[s|t_f, t_i]\,, \tag{2.88a}$$

$$\frac{d}{dt}\left\{ \frac{\hbar}{i} \frac{\delta \tilde{\mathcal{U}}_0\,[s|t_f, t_i]}{\delta K^l(t)} \right\}$$

$$= \int_{-\infty}^{+\infty} d^N x\, d^N y\, b_{l_\alpha}(x,y)\, \Delta_{-\alpha}\left(\frac{\hbar}{i}\frac{\delta}{\delta J(t)} - x\,,\, \frac{\hbar}{i}\frac{\delta}{\delta K(t)} - y \right) \tilde{\mathcal{U}}_0\,[s|t_f, t_i]$$

$$+ J_l(t)\, \tilde{\mathcal{U}}_0\,[s|t_f, t_i]\,, \tag{2.88b}$$

where

$$\Delta_{-\alpha}\left(\frac{\hbar}{i}\frac{\delta}{\delta J(t)}-x,\frac{\hbar}{i}\frac{\delta}{\delta K(t)}-y\right) \equiv (2\pi\hbar)^{-2N}$$
$$\times \lim_{t'\uparrow t}\int_{-\infty}^{+\infty} d^N u\, d^N v \exp\left[\frac{i}{\hbar}\left(\frac{1}{2}-\alpha\right)u\cdot v\right]$$
$$\times \exp\left[\frac{i}{\hbar}\left(x-\frac{\hbar}{i}\frac{\delta}{\delta J(t)}\right)\cdot u\right]\exp\left[\frac{i}{\hbar}\left(y-\frac{\hbar}{i}\frac{\delta}{\delta K(t')}\right)\cdot v\right]. \quad (2.89)$$

The strategy for integrating Schwinger equations (2.88) goes qualitatively as before. We substitute (2.71) into (2.88) and then look for the system of functional differential equations satisfied by $\tilde{\mathcal{U}}_0[\tilde{s}|t_f,t_i]$. It is found to read

$$\frac{i}{\hbar}\left[\dot{q}^l(t) - \Omega^{(a)l}\left(q(t),p(t)\right)\right]\tilde{\mathcal{U}}_0\left[\tilde{s}|t_f,t_i\right] = \frac{\delta\tilde{\mathcal{U}}_0\left[\tilde{s}|t_f,t_i\right]}{\delta p_l(t)}, \quad (2.90a)$$
$$\frac{i}{\hbar}\left[-\dot{p}_l(t) + \Omega_l^{(b)}\left(q(t),p(t)\right)\right]\tilde{\mathcal{U}}_0\left[\tilde{s}|t_f,t_i\right] = \frac{\delta\tilde{\mathcal{U}}_0\left[\tilde{s}|t_f,t_i\right]}{\delta q^l(t)}. \quad (2.90b)$$

Here,

$$\Omega^{(a)l}\left(q(t),p(t)\right)$$
$$\equiv \exp\left[-\frac{i}{\hbar}\frac{q(t)\cdot p(t)}{\left(\frac{1}{2}-\alpha\right)}\right] a_\alpha^l(\phi,\xi)\exp\left[\frac{i}{\hbar}\frac{q(t)\cdot p(t)}{\left(\frac{1}{2}-\alpha\right)}\right], \quad (2.91a)$$
$$\Omega_l^{(b)}\left(q(t),p(t)\right)$$
$$\equiv \exp\left[-\frac{i}{\hbar}\frac{q(t)\cdot p(t)}{\left(\frac{1}{2}-\alpha\right)}\right] b_{l_\alpha}(\phi,\xi)\exp\left[\frac{i}{\hbar}\frac{q(t)\cdot p(t)}{\left(\frac{1}{2}-\alpha\right)}\right], \quad (2.91b)$$

where

$$\phi^k \equiv \frac{\hbar}{i}\left(\frac{1}{2}-\alpha\right)\frac{\partial}{\partial p_k(t)}, \quad (2.92a)$$
$$\xi_k \equiv \frac{\hbar}{i}\left(\frac{1}{2}-\alpha\right)\frac{\partial}{\partial q^k(t)}. \quad (2.92b)$$

To see how this come about we shall start by concentrating on the first term in the right hand side of Eq.(2.88a). From Eqs.(2.71) and (2.89) it follows that

$$\int_{-\infty}^{+\infty} d^N x\, d^N y\, a^l_\alpha(x,y)\, \Delta_{-\alpha}\left(\frac{\hbar}{i}\frac{\delta}{\delta J(t)} - x\,,\, \frac{\hbar}{i}\frac{\delta}{\delta K(t)} - y\right) \tilde{\mathcal{U}}_0\left[s|t_f,t_i\right]$$
$$= \int_{-\infty}^{+\infty} d^N q_f \int_{-\infty}^{+\infty} d^N q_i\, \psi_0^*(q_f)\, \psi_0(q_i) \int [\mathcal{D}q] \int [\mathcal{D}p]\, \tilde{\mathcal{U}}_0[\tilde{s}|t_f,t_i]$$
$$\times \int_{-\infty}^{+\infty} d^N x\, d^N y\, a^l_\alpha(x,y)\, (2\pi\hbar)^{-2N} \lim_{t'\uparrow t} \int_{-\infty}^{+\infty} d^N u\, d^N v$$
$$\times \exp\left[\frac{i}{\hbar}\left(\frac{1}{2}-\alpha\right) u\cdot v\right] \exp\left[\frac{i}{\hbar}\left(x^i - q^i(t)\right)\cdot u_i(t)\right] \exp\left[\frac{i}{\hbar}\left(y_k - p_k(t')\right)\cdot v^k\right]$$
$$\times \exp\left\{\frac{i}{\hbar}\int_{t_i}^{t_f} \sum_{j=1}^{N}\left[q^j(\tau)J_j(\tau) + p_j(\tau)K^j(\tau)\right]\right\}. \tag{2.93}$$

The u and v integrals are straightforward; after carrying them out, we find

$$\int_{-\infty}^{+\infty} d^N x\, d^N y\, a^l_\alpha(x,y)\, \Delta_{-\alpha}\left(\frac{\hbar}{i}\frac{\delta}{\delta J(t)} - x\,,\, \frac{\hbar}{i}\frac{\delta}{\delta K(t)} - y\right) \tilde{\mathcal{U}}_0\left[s|t_f,t_i\right]$$
$$= \int_{-\infty}^{+\infty} d^N q_f \int_{-\infty}^{+\infty} d^N q_i\, \psi_0^*(q_f)\, \psi_0(q_i) \int [\mathcal{D}q] \int [\mathcal{D}p]\, \tilde{\mathcal{U}}_0[\tilde{s}|t_f,t_i]$$
$$\times \int_{-\infty}^{+\infty} d^N x\, d^N y\, a^l_\alpha(x,y)\, \frac{(2\pi\hbar)^{-N}}{\left(\frac{1}{2}-\alpha\right)^N}$$
$$\times \exp\left\{-\frac{i}{\hbar}\left[y_k - p_k(t)\right]\frac{1}{\left(\frac{1}{2}-\alpha\right)}\left[x^k - q^k(t)\right]\right\}$$
$$\times \exp\left\{\frac{i}{\hbar}\int_{t_i}^{t_f} \sum_{j=1}^{N}\left[q^j(\tau)J_j(\tau) + p_j(\tau)K^j(\tau)\right]\right\}, \tag{2.94}$$

where the limit $t' \uparrow t$ was already taken. Now

$$a^l_\alpha(x,y)\, \exp\left\{-\frac{i}{\hbar}\left[y_k - p_k(t)\right]\frac{1}{\left(\frac{1}{2}-\alpha\right)}\left[x^k - q^k(t)\right]\right\}$$
$$= \exp\left[-\frac{i}{\hbar}\frac{p\cdot q}{\left(\frac{1}{2}-\alpha\right)}\right] a^l_\alpha\left[\frac{\hbar}{i}\left(\frac{1}{2}-\alpha\right)\frac{\partial}{\partial p}\,,\, \frac{\hbar}{i}\left(\frac{1}{2}-\alpha\right)\frac{\partial}{\partial q}\right]$$
$$\times \exp\left[\frac{i}{\hbar}\frac{\left[y\cdot q - y\cdot x + p\cdot x\right]}{\left(\frac{1}{2}-\alpha\right)}\right]. \tag{2.95}$$

When this result is fed back into Eq.(2.94) and the integrations on x and y are performed, we end up with

$$\begin{aligned}
&\int_{-\infty}^{+\infty} d^N x\, d^N y\, a^l_\alpha(x,y)\, \Delta_{-\alpha}\left(\frac{\hbar}{i}\frac{\delta}{\delta J(t)} - x\,,\, \frac{\hbar}{i}\frac{\delta}{\delta K(t)} - y\right) \tilde{\mathcal{U}}_0\left[s|t_f,t_i\right] \\
&= \int_{-\infty}^{+\infty} d^N q_f \int_{-\infty}^{+\infty} d^N q_i\, \psi_0^*(q_f)\, \psi_0(q_i) \\
&\times \int [\mathcal{D}q] \int [\mathcal{D}p]\, \tilde{\mathcal{U}}_0[\tilde{s}|t_f,t_i]\, \Omega^{(a)l}\left(q(t),p(t)\right) \\
&\times \exp\left\{\frac{i}{\hbar}\int_{t_i}^{t_f} \sum_{j=1}^{N} \left[q^j(\tau)J_j(\tau) + p_j(\tau)K^j(\tau)\right]\right\}
\end{aligned} \tag{2.96}$$

in agreement with Eqs.(2.90a) and (2.91a). A similar analysis corroborates Eqs.(2.90b) and (2.91b).

This approach becomes useful after learning how to compute $a^l_\alpha(q,p)$ and $b_{l_\alpha}(q,p)$. We conjecture that

$$a^l_\alpha(q,p) = \frac{\partial h_\alpha(q,p)}{\partial p_l}\,, \tag{2.97a}$$

$$b_{l_\alpha}(q,p) = -\frac{\partial h_\alpha(q,p)}{\partial q^l}\,, \tag{2.97b}$$

where $h_\alpha(q,p)$ is the GWT of the Hamiltonian operator. To substantiate Eq.(2.97a) we start by combining Eqs.(1.52) and (2.52a). It gives

$$\begin{aligned}
&a^l_\alpha(q,p) \\
&= \int_{-\infty}^{+\infty} d^N v e^{\frac{i}{\hbar}p\cdot v} \left\langle q - \left(\frac{1}{2}-\alpha\right)v\right| A^l(Q,P) \left|q + \left(\frac{1}{2}+\alpha\right)v\right\rangle \\
&= \frac{i}{\hbar}\int_{-\infty}^{+\infty} d^N v e^{\frac{i}{\hbar}p\cdot v} \\
&\times \left\langle q - \left(\frac{1}{2}-\alpha\right)v\right| \left[H, Q^l\right] \left|q + \left(\frac{1}{2}+\alpha\right)v\right\rangle .
\end{aligned} \tag{2.98}$$

Now

$$\begin{aligned}
&\left\langle q-\left(\frac{1}{2}-\alpha\right)v\right|\left[H,Q^l\right]\left|q+\left(\frac{1}{2}+\alpha\right)v\right\rangle \\
&=\left\langle q-\left(\frac{1}{2}-\alpha\right)v\right|H\,Q^l\left|q+\left(\frac{1}{2}+\alpha\right)v\right\rangle \\
&-\left\langle q-\left(\frac{1}{2}-\alpha\right)v\right|Q^l\,H\left|q+\left(\frac{1}{2}+\alpha\right)v\right\rangle \\
&=v^l\left\langle q-\left(\frac{1}{2}-\alpha\right)v\right|H\left|q+\left(\frac{1}{2}+\alpha\right)v\right\rangle . \qquad (2.99)
\end{aligned}$$

Hence, Eqs.(2.98) and (2.99) yield

$$\begin{aligned}
a^l_\alpha(q,p) &= \int_{-\infty}^{+\infty} d^N v\left(\frac{i}{\hbar}v^l\right)e^{\frac{i}{\hbar}p\cdot v} \\
&\times\left\langle q-\left(\frac{1}{2}-\alpha\right)v\right|H(Q,P)\left|q+\left(\frac{1}{2}+\alpha\right)v\right\rangle \\
&=\frac{\partial h_\alpha(q,p)}{\partial p_l} \qquad (2.100)
\end{aligned}$$

in agreement with Eq.(2.97a) (*QED*). As for Eq.(2.97b), the point of departure is

$$\begin{aligned}
&b_{l_\alpha}(q,p) \\
&=\int_{-\infty}^{+\infty} d^N v e^{\frac{i}{\hbar}p\cdot v}\left\langle q-\left(\frac{1}{2}-\alpha\right)v\right|B_l(Q,P)\left|q+\left(\frac{1}{2}+\alpha\right)v\right\rangle \\
&=\frac{i}{\hbar}\int_{-\infty}^{+\infty} d^N v e^{\frac{i}{\hbar}p\cdot v} \\
&\times\left\langle q-\left(\frac{1}{2}-\alpha\right)v\right|\left[H,P_l\right]\left|q+\left(\frac{1}{2}+\alpha\right)v\right\rangle , \qquad (2.101)
\end{aligned}$$

as it can be checked by employing Eqs.(1.52) and (2.52b). Then

$$
\begin{aligned}
& b_{l_\alpha}(q,p) \\
&= \frac{i}{\hbar}\int_{-\infty}^{+\infty} d^N v e^{\frac{i}{\hbar}p\cdot v}\int_{-\infty}^{+\infty} d^N q' \\
&\times \left[\langle q - \left(\frac{1}{2}-\alpha\right) v|H|q'\rangle\langle q'|P_l|q+\left(\frac{1}{2}+\alpha\right) v\rangle \right. \\
&- \left. \langle q - \left(\frac{1}{2}-\alpha\right) v|P_l|q'\rangle\langle q'|H|q+\left(\frac{1}{2}+\alpha\right) v\rangle \right] \\
&= \int_{-\infty}^{+\infty} d^N v e^{\frac{i}{\hbar}p\cdot v}\int_{-\infty}^{+\infty} d^N q' \\
&\times \left[\langle q - \left(\frac{1}{2}-\alpha\right) v|H|q'\rangle \frac{\partial}{\partial q'^{\,l}}\delta(q'-q_1) \right. \\
& \left. +\langle q'|H|q+\left(\frac{1}{2}+\alpha\right) v\rangle \frac{\partial}{\partial q'^{\,l}}\delta(q'-q_2)\right] \\
&= -\int_{-\infty}^{+\infty} d^N v e^{\frac{i}{\hbar}p\cdot v} \\
&\times \left[\left(\frac{\partial}{\partial q_1^l}+\frac{\partial}{\partial q_2^l}\right)\langle q - \left(\frac{1}{2}-\alpha\right) v|H|q+\left(\frac{1}{2}+\alpha\right) v\rangle\right], \qquad (2.102)
\end{aligned}
$$

where

$$q_1^l \equiv q^l + \left(\frac{1}{2}+\alpha\right) v^l \,, \qquad (2.103a)$$

$$q_2^l \equiv q^l - \left(\frac{1}{2}-\alpha\right) v^l \,. \qquad (2.103b)$$

By solving this last set of equations for q and v in terms of q_1 and q_2 we obtain

$$q^l = \left(\frac{1}{2}-\alpha\right) q_1^l + \left(\frac{1}{2}+\alpha\right) q_2^l \,, \qquad (2.104a)$$

$$v^l = q_1^l - q_2^l \,. \qquad (2.104b)$$

In turns, the chain rule for differentiation yields

$$\frac{\partial}{\partial q_1^l} = \frac{\partial q^k}{\partial q_1^l}\frac{\partial}{\partial q^k} + \frac{\partial v^k}{\partial q_1^l}\frac{\partial}{\partial v^k} = \left(\frac{1}{2}-\alpha\right)\frac{\partial}{\partial q^l} + \frac{\partial}{\partial v^l}\,, \qquad (2.105a)$$

$$\frac{\partial}{\partial q_2^l} = \frac{\partial q^k}{\partial q_2^l}\frac{\partial}{\partial q^k} + \frac{\partial v^k}{\partial q_2^l}\frac{\partial}{\partial v^k} = \left(\frac{1}{2}+\alpha\right)\frac{\partial}{\partial q^l} - \frac{\partial}{\partial v^l}\,, \qquad (2.105b)$$

implying that

$$\frac{\partial}{\partial q_1^l} + \frac{\partial}{\partial q_2^l} = \frac{\partial}{\partial q^l} \,. \tag{2.106}$$

Hence, by coming back with Eq.(2.106) into Eq.(2.102) we obtain

$$\begin{aligned} & b_{l_\alpha}(q,p) \\ & = -\frac{\partial}{\partial q^l} \int_{-\infty}^{+\infty} d^N v e^{\frac{i}{\hbar} p \cdot v} \langle q - \left(\frac{1}{2} - \alpha\right) v | H | q + \left(\frac{1}{2} + \alpha\right) v \rangle \\ & = -\frac{\partial}{\partial q^l} h_\alpha(q,p) \end{aligned} \tag{2.107}$$

as proposed in Eq.(2.97b) (*QED*).

We shall stop for a while to check that Eqs.(2.90) reduce to (2.78) for the one dimensional system whose Hamiltonian operator is that given in Eq.(2.53). To verify this we recall that, according to the developments in subsection 1.2.2, the corresponding GWT does not depend on α and it can be written as

$$h(q,p) = f(q) + g(p) \,, \tag{2.108}$$

where $f(q)$ and $g(p)$ are the GWT of $F(Q)$ and $G(P)$, respectively. Therefore, (see Eqs.(2.97))

$$a_\alpha(q,p) = a(q,p) = \frac{\partial h(q,p)}{\partial p} = \frac{\partial g(p)}{\partial p} =: g'(p) \,, \tag{2.109a}$$

$$b_\alpha(q,p) = a(q,p) = -\frac{\partial h(q,p)}{\partial q} = -\frac{\partial f(q)}{\partial q} =: -f'(q) \,. \tag{2.109b}$$

Correspondingly,

$$a_\alpha(\phi,\xi) = g'(\xi) \,, \tag{2.110a}$$

$$b_\alpha(\phi,\xi) = -f'(\phi) \,. \tag{2.110b}$$

As for Eqs.(2.91), they boil down to

$$\begin{aligned}
&\Omega^{(a)}\left(q(t),p(t)\right) \\
&= \exp\left[-\frac{i}{\hbar}\frac{q(t)p(t)}{\left(\frac{1}{2}-\alpha\right)}\right] g'(\xi)\, \exp\left[\frac{i}{\hbar}\frac{q(t)p(t)}{\left(\frac{1}{2}-\alpha\right)}\right] \\
&= g'(p)\,, \qquad (2.111a)\\
&\Omega^{(b)}\left(q(t),p(t)\right) \\
&= -\exp\left[-\frac{i}{\hbar}\frac{q(t)p(t)}{\left(\frac{1}{2}-\alpha\right)}\right] f'(\phi)\, \exp\left[\frac{i}{\hbar}\frac{q(t)p(t)}{\left(\frac{1}{2}-\alpha\right)}\right] \\
&= -f'(q)\,, \qquad (2.111b)
\end{aligned}$$

which after being replaced into Eqs.(2.90) lead us back to Eqs.(2.78) (*QED*).

We shall now proceed by elaborating on the α-dependence of Schwinger equations (2.90) in the general case. Once again, we restrict ourselves to deal with examples. First, we shall bring again into consideration the case of the modified one dimensional harmonic oscillator whose dynamics is specified by the Hamiltonian operator in Eq.(1.229). From the corresponding GWT, quoted in Eq.(1.230), it follows that

$$\begin{aligned}
&a_\alpha(q,p) = \frac{p}{M} + \omega q \Longrightarrow a_\alpha(\phi,\xi) = \frac{\hbar}{i}\left(\frac{1}{2}-\alpha\right) \\
&\times \left(\omega\frac{\partial}{\partial p} + \frac{1}{M}\frac{\partial}{\partial q}\right), \qquad (2.112a)\\
&b_\alpha(q,p) = -2M\omega^2 q - \omega p \Longrightarrow b_\alpha(\phi,\xi) = \frac{\hbar}{i}\left(\frac{1}{2}-\alpha\right) \\
&\times \left(-2M\omega^2\frac{\partial}{\partial p} - \omega\frac{\partial}{\partial q}\right), \qquad (2.112b)
\end{aligned}$$

in accordance with Eqs.(2.92). The coefficient functions in Schwinger equations (2.90) can be computed at once by employing (2.91). The results turn out to be independent of α and read

$$\Omega^{(a)}\left(q,p\right) = \omega q + \frac{p}{M}\,, \qquad (2.113a)$$

$$\Omega^{(b)}\left(q,p\right) = -2M\omega q - \omega p\,. \qquad (2.113b)$$

This, in turns, secures that $\tilde{\mathcal{U}}_0\left[\tilde{s}|t_f,t_i\right]$ (see Eqs.(2.90)) does not depend on α.

A less trivial one dimensional model arises from the Hamiltonian

$$H(Q,P) = \frac{QP^2Q}{ML^2}, \tag{2.114}$$

where L is a constant with dimensions of length. By using Eq.(1.52) we find

$$h_\alpha(q,p) = \frac{q^2p^2}{ML^2} + 4\left(\frac{\hbar}{i}\right)\alpha\frac{qp}{ML^2} - \left(\frac{\hbar}{i}\right)^2\frac{1}{ML^2}\left(\frac{1}{2} - 2\alpha^2\right) \tag{2.115}$$

and, correspondingly,

$$a_\alpha(\phi,\xi) = \frac{2}{ML^2}\left(\frac{\hbar}{i}\right)^3\left(\frac{1}{2}-\alpha\right)^3\frac{\partial}{\partial q}\frac{\partial^2}{\partial p^2}$$
$$+\frac{4}{ML^2}\left(\frac{\hbar}{i}\right)^2\left(\frac{1}{2}-\alpha\right)\alpha\frac{\partial}{\partial p}, \tag{2.116a}$$
$$b_\alpha(\phi,\xi) = -\frac{2}{ML^2}\left(\frac{\hbar}{i}\right)^3\left(\frac{1}{2}-\alpha\right)^3\frac{\partial}{\partial p}\frac{\partial^2}{\partial q^2}$$
$$-\frac{4}{ML^2}\left(\frac{\hbar}{i}\right)^2\left(\frac{1}{2}-\alpha\right)\alpha\frac{\partial}{\partial q}. \tag{2.116b}$$

Hence, by using Eqs.(2.91) we get

$$\Omega^{(a)}(q,p) = \frac{2}{ML^2}q^2p + \frac{2}{ML^2}\left(\frac{\hbar}{i}q\right), \tag{2.117a}$$
$$\Omega^{(b)}(q,p) = -\frac{2}{ML^2}qp^2 - \frac{2}{ML^2}\left(\frac{\hbar}{i}p\right), \tag{2.117b}$$

which do not depend on α. Because of the reasons given above this property also applies for $\tilde{\mathcal{U}}_0\left[\tilde{s}|t_f,t_i\right]$.

We are not aware of a general proof ensuring the independence on α of $\tilde{\mathcal{U}}_0\left[s|t_f,t_i\right]$.

Chapter 3

Generating functional of Green functions

The generating functional of Green functions $\tilde{\mathcal{U}}_0\,[s|t_f, t_i]$ is the cornerstone of the functional formulation of quantum mechanics. For realistic models its exact evaluation is hopeless. The harmonic oscillator represents an exception to this rule. We consider this model in detail for pedagogical and technical reasons; it will help us to develop a systematics of general validity. Afterwards, the functional framework is shown to harmonize with the canonical algebra. The peculiarities of systems with a continuous energy spectrum is closely examined. The form assumed by $\tilde{\mathcal{U}}_0\,[s|t_f, t_i]$ at the limit $t_f \to +\infty$, $t_i \to -\infty$ is derived in full generality. We then look for the expansion of $\tilde{W}_0\,[s|+\infty, -\infty]$ in powers of $\hbar$. The effective action and the effective potential are also discussed.

3.1 Generating functional for the one dimensional harmonic oscillator

Our starting point is $\tilde{\mathcal{U}}_0\,[\tilde{s}|t_f, t_i]$, given in Eq.(2.79), when specialized for $N = 1$. Also

$$h(q, p) \;=\; \frac{p^2}{2M} \;+\; \frac{1}{2}\,M\omega^2 q^2\,. \tag{3.1}$$

Therefore,

$$\tilde{\mathcal{U}}_0\,[\tilde{s}|t_f, t_i] \;=\; \mathcal{N}\,\exp\left\{\frac{i}{\hbar}\int_{t_i}^{t_f} dt\,\left[p(t)\dot{q}(t) \;-\; \frac{p^2}{2M} \;-\; \frac{1}{2}\,M\omega^2 q^2\right]\right\}. \tag{3.2}$$

In accordance with Eq.(2.82) we have that

$$
\begin{aligned}
{}_H\langle q_f, t_f | q_i, t_i \rangle_H^J &= \mathcal{N} \int [\mathcal{D}q] \int [\mathcal{D}p] \\
&\times \exp \left\{ \frac{i}{\hbar} \int_{t_i}^{t_f} dt \left[p(t)\dot{q}(t) - \frac{p^2}{2M} - \frac{1}{2} M\omega^2 q^2 \right] \right. \\
&\left. + \frac{i}{\hbar} \int_{t_i}^{t_f} dt\, q(t) J(t) \right\} . \qquad (3.3)
\end{aligned}
$$

In Chapter 1 we computed ${}_H\langle q_f, t_f | q_i, t_i \rangle_H^{J=0}$ by using the definition of the phase space path integral based on the time slicing procedure. We must now decide about the definition to be used when the fictitious source J is *turned on*. As a matter of simplicity we give up that based on the time slicing procedure and then choose the one presented and discussed in appendix B. Of course, this change should not affect the results.

The integral on p looks formally like that in Eq.(B.48). However, the collection of functions $p(t)$ defining the domain of integration is not restricted by any constraint. Indeed, $p(t_i)$ and $p(t_f)$ fluctuate between $\pm\infty$. However, we shall assume that this integral can still be performed by invoking the translational invariance of the functional integral. Accordingly, we find [1]

$$
{}_H\langle q_f, t_f | q_i, t_i \rangle_H^J = \mathcal{N} \int [\mathcal{D}q] \, e^{\frac{i}{\hbar} S^J[q]} , \qquad (3.4)
$$

where

$$
S^J[q] = \int_{t_i}^{t_f} dt \left(\frac{1}{2} M \dot{q}^2 - \frac{1}{2} M\omega^2 q^2 + q J \right) . \qquad (3.5)
$$

The integration domain in the right hand side of Eq.(3.4) is defined by the *inhomogeneous* boundary conditions

$$
q(t_i) = q_i , \qquad q(t_f) = q_f . \qquad (3.6)
$$

Therefore, to call upon the techniques of integration developed in Appendix B we must first change the variable of integration in Eq.(3.4) as follows [Feynman and Hibbs (1965)]

$$
q(t) = q_{cl}(t) + \eta(t) . \qquad (3.7)
$$

[1] All constants are being absorbed into the normalization constant.

Here, $q_{cl}(t)$ is a solution of the equations of motion

$$\left.\frac{\delta S^J[q]}{\delta q(t)}\right|_{q=q_{cl}} = 0 \tag{3.8}$$

verifying the boundary conditions

$$q_{cl}(t_i) = q_i\,, \qquad q_{cl}(t_f) = q_f\,. \tag{3.9}$$

The developments in Appendix B now definitely apply since the new domain of integration is defined by the following homogeneous boundary conditions

$$\eta(t_i) = 0\,, \qquad \eta(t_f) = 0\,. \tag{3.10}$$

The present methodology is part of a more general scheme known as *steepest descent* or *stationary phase approximation*.

We shall next write the functional integral in Eq.(3.4) in terms of $\eta(t)$. From the functional point of view $q_{cl}(t)$ acts as a *constant*. Hence, as far as the integration measure is concerned we have that

$$[\mathcal{D}q] = [\mathcal{D}\eta]\,. \tag{3.11}$$

On the other hand, after replacing Eq.(3.7) into Eq.(3.5) we find

$$\begin{aligned} S^J[q] &= S^J[q_{cl}] + \int_{t_i}^{t_f} dt\,\eta(t)\left(-M\ddot{q}_{cl} - M\omega^2 q_{cl} + J\right) \\ &+ \frac{M}{2}\int_{t_i}^{t_f} dt\,\eta(t)\left(-\frac{d^2}{dt^2} - \omega^2\right)\eta(t)\,, \end{aligned} \tag{3.12}$$

where the second and third terms emerge after by part integrations. The correspondent surface terms drop out as a consequence of the boundary conditions in Eq.(3.10). We incorporate below the information that $q_{cl}(t)$ fulfils the equation of motion

$$M\ddot{q}_{cl} + M\omega^2 q_{cl} - J = 0\,, \tag{3.13}$$

which arises after substituting Eq.(3.5) into Eq.(3.8). Hence, the second term in the right hand side of Eq.(3.12) disappears altogether and the just mentioned equation reduces to

$$S^J[q] = S^J[q_{cl}] + \frac{1}{2}\int_{t_i}^{t_f} dt \int_{t_i}^{t_f} dt'\,\eta(t)\,\Xi(t,t')\,\eta(t')\,, \tag{3.14}$$

where $\Xi(t,t')$ is the local operator

$$\Xi(t,t') \equiv M\left(-\frac{d^2}{dt^2} - \omega^2\right) \delta(t-t') \,. \tag{3.15}$$

By replacing Eqs.(3.11) and (3.14) into Eq.(3.4) and after using Eq.(B.66) we arrive at

$$\begin{aligned} &{}_H\langle q_f, t_f | q_i, t_i \rangle_H^J \\ &= \mathcal{N} e^{\frac{i}{\hbar} S^J[q_{cl}]} \int [\mathcal{D}\eta]\, e^{\frac{i}{2\hbar}\int_{t_i}^{t_f} dt \int_{t_i}^{t_f} dt'\, \eta(t)\, \Xi(t,t')\, \eta(t')} \\ &= \mathcal{N} \left[\det\left(\frac{\Xi}{\hbar}\right)\right]^{-\frac{1}{2}} e^{\frac{i}{\hbar} S^J[q_{cl}]} \,. \end{aligned} \tag{3.16}$$

The primary object of interest, $\tilde{\mathcal{U}}_0\,[J|t_f,t_i]$, is related to the propagator through the following (see Eq.(2.68))

$$\begin{aligned} &\tilde{\mathcal{U}}_0\,[J|t_f,t_i] \\ &= \int_{-\infty}^{+\infty} dq_f \int_{-\infty}^{+\infty} dq_i\, \psi_0^*(q_f)\, {}_H\langle q_f, t_f | q_i, t_i \rangle_H^J\, \psi_0(q_i) \,, \end{aligned} \tag{3.17}$$

where

$$\psi_0(q) = \left(\frac{M\omega}{\pi\hbar}\right)^{\frac{1}{4}} e^{-\frac{M\omega}{2\hbar} q^2} \tag{3.18}$$

is the ground state eigenfunction for the one dimensional harmonic oscillator.

Finding $\tilde{\mathcal{U}}_0\,[J|t_f,t_i]$ demands to carry out the integrals in Eq.(3.17). We turn, then, to uncovering the dependence of $S^J[q_{cl}]$ on q_f and q_i. This information is buried in the solution of the classical equation of motion Eq.(3.13) which can be written as

$$q_{cl}(t) = q_h(t) + q_p(t) \,. \tag{3.19}$$

Here, $q_h(t)$ solves the ordinary homogeneous differential equation

$$\ddot{q}_h + \omega^2 q_h = 0 \,, \tag{3.20}$$

while $q_p(t)$ solves the particular equation

$$\left(\frac{d^2}{dt^2} + \omega^2\right) q_p = \frac{1}{M} J \,. \tag{3.21}$$

The boundary conditions in Eq.(3.9) are incorporated by demanding

$$q_h(t_i) = q_i \,, \qquad q_h(t_f) = q_f \,, \tag{3.22a}$$

$$q_p(t_i) = 0 \,, \qquad q_h(t_f) = 0 \,. \tag{3.22b}$$

We find that

$$q_h(t) = \frac{1}{\sin\omega\,(t_f - t_i)} \left[q_f \sin\omega(t - t_i) - q_i \sin\omega(t - t_f)\right] \,, \tag{3.23}$$

whereas

$$q_p(t) = \frac{1}{M} \int_{t_i}^{t_f} dt' \,\Delta(t - t')\, J(t') \,. \tag{3.24}$$

Here, $\Delta(t-t')$ is the Green function of the local operator $\left(\frac{d^2}{dt^2} + \omega^2\right) \delta(t-t')$, i.e.,

$$\left(\frac{d^2}{dt^2} + \omega^2\right) \Delta(t - t') = \delta(t - t') \,. \tag{3.25}$$

The invariance of the operator under time translations dictates the form of the dependence of Δ on t and t', i.e., $\Delta(t - t')$. Furthermore, the boundary conditions in Eq.(3.22b) demand that Eq.(3.25) be solved under the following restrictions

$$\Delta(t_f - t') = \Delta(t_i - t') = 0 \,. \tag{3.26}$$

This solution reads

$$\begin{aligned} \Delta(t - t') = &- \frac{1}{\omega \sin\omega\,(t_f - t_i)} \left[\theta(t' - t) \sin\omega\,(t - t_i) \sin\omega\,(t_f - t')\right. \\ &+ \left.\theta(t - t') \sin\omega\,(t' - t_i) \sin\omega\,(t_f - t)\right] \,. \end{aligned} \tag{3.27}$$

Next, we substitute the just found solution for $q_{cl}(t)$ within the functional argument of $S^J[q_{cl}]$. By starting from Eq.(3.5) we get

$$\begin{aligned} S^J[q_{cl}] &= \int_{t_i}^{t_f} dt \left(\frac{1}{2} M \dot{q}_{cl}^2 - \frac{1}{2} M\omega^2 q_{cl}^2 + q_{cl} J \right) \\ &= \frac{M}{2} [q_f \dot{q}_{cl}(t_f) - q_i \dot{q}_{cl}(t_i)] \\ &- \frac{M}{2} \int_{t_i}^{t_f} dt q_{cl}(t) \left(\ddot{q}_{cl} + \omega^2 q_{cl} - \frac{2}{M} J \right) \\ &= \frac{M}{2} [q_f \dot{q}_{cl}(t_f) - q_i \dot{q}_{cl}(t_i)] + \frac{1}{2} \int_{t_i}^{t_f} dt q_{cl}(t) J(t) \,. \end{aligned} \tag{3.28}$$

The third term in the latter equality results from the second after a by part integration of the factor containing $\dot{q}_{cl}^2$. Afterwards, the equation of motion (3.13) was used for arriving at the algebraic form exhibited by the last term. Furthermore, the splitting in Eq.(3.19) implies that

$$\begin{aligned} \frac{M}{2} [q_f \dot{q}_{cl}(t_f) - q_i \dot{q}_{cl}(t_i)] &= \frac{M}{2} [q_f \dot{q}_h(t_f) - q_i \dot{q}_h(t_i)] \\ &+ \frac{M}{2} [q_f \dot{q}_p(t_f) - q_i \dot{q}_p(t_i)] \,, \end{aligned} \tag{3.29a}$$

$$\frac{1}{2} \int_{t_i}^{t_f} dt q_{cl}(t) J(t) = \frac{1}{2} \int_{t_i}^{t_f} dt q_h(t) J(t) + \frac{1}{2} \int_{t_i}^{t_f} dt q_p(t) J(t). \tag{3.29b}$$

By taking into account the explicit forms of $q_h(t)$ and $q_p(t)$ in Eqs.(3.23) and (3.24) we obtain

$$\begin{aligned} \frac{M}{2} [q_f \dot{q}_h(t_f) - q_i \dot{q}_h(t_i)] &= \frac{M\omega}{2 \sin \omega(t_f - t_i)} \\ &\times \left[\left(q_f^2 + q_i^2 \right) \cos \omega(t_f - t_i) - 2 q_f q_i \right] \,, \end{aligned} \tag{3.30}$$

$$\begin{aligned} \frac{1}{2} \int_{t_i}^{t_f} dt q_h(t) J(t) &= \frac{1}{2 \sin \omega(t_f - t_i)} \left[q_f \int_{t_i}^{t_f} dt \sin \omega(t - t_i) J(t) \right. \\ &\left. + q_i \int_{t_i}^{t_f} dt \sin \omega(t_f - t) J(t) \right] \,, \end{aligned} \tag{3.31}$$

$$\begin{aligned} \frac{M}{2} [q_f \dot{q}_p(t_f) - q_i \dot{q}_p(t_i)] &= \frac{1}{2 \sin \omega(t_f - t_i)} \left[q_f \int_{t_i}^{t_f} dt \sin \omega(t - t_i) J(t) \right. \\ &\left. + q_i \int_{t_i}^{t_f} dt \sin \omega(t_f - t) J(t) \right] \,, \end{aligned} \tag{3.32}$$

and

$$\frac{1}{2}\int_{t_i}^{t_f} dt q_p(t)J(t) = -\frac{1}{M\omega\sin\omega(t_f-t_i)}$$
$$\times \int_{t_i}^{t_f} dt \int_{t_i}^{t_f} dt' J(t)\theta(t-t')\sin\omega(t_f-t)\sin\omega(t'-t_i)J(t'). \quad (3.33)$$

By putting everything back together into Eq.(3.28) we achieve the final form [Feynman (1950)]

$$S^J[q_{cl}] = \frac{M\omega}{2\sin\omega(t_f-t_i)}\Big[\left(q_f^2+q_i^2\right)\cos\omega(t_f-t_i) - 2q_f q_i$$
$$+\frac{2q_f}{M\omega}\int_{t_i}^{t_f} dt\sin\omega(t-t_i)J(t) + \frac{2q_i}{M\omega}\int_{t_i}^{t_f} dt\sin\omega(t_f-t)J(t) - \frac{2}{M^2\omega^2}$$
$$\times \int_{t_i}^{t_f} dt \int_{t_i}^{t_f} dt' J(t)\,\theta(t-t')\sin\omega(t_f-t)\sin\omega(t'-t_i)J(t')\Big]. \quad (3.34)$$

We return next with Eq.(3.34) into (3.16). The result thus obtained, along with Eq.(3.18), is brought into Eq.(3.17). The outcome is

$$\tilde{\mathcal{U}}_0[J|t_f,t_i] = \mathcal{N}\left[\det\left(\tfrac{\Xi}{\hbar}\right)\right]^{-\frac{1}{2}}\left(\frac{M\omega}{\pi\hbar}\right)^{\frac{1}{2}}\exp\Big[-\frac{i}{\hbar M\omega\sin\omega T}$$
$$\times \int_{t_i}^{t_f} dt \int_{t_i}^{t_f} dt' J(t)\,\theta(t-t')\sin\omega(t_f-t)\sin\omega(t'-t_i)J(t')\Big]$$
$$\times \Lambda[J|t_f,t_i], \quad (3.35)$$

where [2]

$$\Lambda[J|t_f,t_i]$$
$$\equiv \int_{-\infty}^{+\infty} dq_f \int_{-\infty}^{+\infty} dq_i \exp\left(-aq_f^2 - aq_i^2 + bq_f q_i + cq_f + dq_i\right), \quad (3.36)$$

and

$$a \equiv \frac{M\omega}{2\hbar}(1 - i\cot\omega T), \quad (3.37a)$$
$$b \equiv -\frac{iM\omega}{\hbar}\frac{1}{\sin\omega T}, \quad (3.37b)$$
$$c \equiv \frac{i}{\hbar}\frac{\int_{t_i}^{t_f} dt\sin\omega(t-t_i)J(t)}{\sin\omega T}, \quad (3.37c)$$
$$d \equiv \frac{i}{\hbar}\frac{\int_{t_i}^{t_f} dt\sin\omega(t_f-t)J(t)}{\sin\omega T}. \quad (3.37d)$$

[2]We remind the reader that $T \equiv t_f - t_i$.

The integrals in Eq.(3.36) can be exactly computed. To that end, we change the integration variables ($q_i, q_f \to x, y$) as follows

$$q_f = \frac{1}{\sqrt{2}}\,x - \frac{1}{\sqrt{2}}\,y\,, \tag{3.38a}$$

$$q_i = \frac{1}{\sqrt{2}}\,x + \frac{1}{\sqrt{2}}\,y\,. \tag{3.38b}$$

The Jacobian of this transformation is 1. Moreover,

$$q_f^2 + q_i^2 = x^2 + y^2 \tag{3.39}$$

and

$$q_f q_i = \frac{1}{2}\left(x^2 - y^2\right)\,. \tag{3.40}$$

Hence, Eq.(3.36) splits as

$$\Lambda[J|t_f, t_i] = \Lambda_1[J|t_f, t_i]\,\Lambda_2[J|t_f, t_i]\,, \tag{3.41}$$

where

$$\Lambda_1[J|t_f, t_i] \equiv \int_{-\infty}^{+\infty} dx \exp\left[-\left(a - \frac{b}{2}\right)x^2 + \frac{1}{\sqrt{2}}\,(c+d)\,x\right], \tag{3.42a}$$

$$\Lambda_2[J|t_f, t_i] \equiv \int_{-\infty}^{+\infty} dy \exp\left[-\left(a + \frac{b}{2}\right)y^2 - \frac{1}{\sqrt{2}}\,(c-d)\,y\right]. \tag{3.42b}$$

By recalling the results in Appendix A we get

$$\Lambda_1[J|t_f, t_i] = \sqrt{\frac{\pi}{a - \frac{b}{2}}}\,\exp\left[\frac{(c+d)^2}{8\left(a - \frac{b}{2}\right)}\right], \tag{3.43a}$$

$$\Lambda_2[J|t_f, t_i] = \sqrt{\frac{\pi}{a + \frac{b}{2}}}\,\exp\left[\frac{(d-c)^2}{8\left(a + \frac{b}{2}\right)}\right], \tag{3.43b}$$

which, in accordance with Eq.(3.41), lead us to

$$\Lambda[J|t_f, t_i] = \frac{2\pi}{\sqrt{4a^2 - b^2}}\,\exp\left[\frac{a}{4a^2 - b^2}\left(d^2 + \frac{bdc}{a} + c^2\right)\right]. \tag{3.44}$$

Once the definitions in Eqs.(3.37) are incorporated into Eq.(3.44) and the corresponding outcome is fed back into Eq.(3.35) we arrive at [3]

$$
\begin{aligned}
\tilde{\mathcal{U}}_0[J|t_f,t_i] &= \exp\Bigg\{ -\frac{1}{2M\omega\hbar\sin^2\omega T}\int_{t_i}^{t_f} dt \int_{t_i}^{t_f} dt'\theta(t-t')J(t) \\
&\times \left[\sin\omega(t_f-t)\sin\omega(t_f-t') + \sin\omega(t-t_i)\sin\omega(t'-t_i)\right. \\
&+ \cos\omega T \sin\omega(t_f-t)\sin\omega(t'-t_i) \\
&+ \left.\cos\omega T\sin\omega(t_f-t')\sin\omega(t-t_i)\right] J(t') + \frac{i}{2M\omega\hbar\sin\omega T} \\
&\times \int_{t_i}^{t_f} dt \int_{t_i}^{t_f} dt'\theta(t-t')J(t)\left[\sin\omega(t_f-t')\sin\omega(t-t_i)\right. \\
&- \left.\sin\omega(t_f-t)\sin\omega(t'-t_i)\right] J(t')\Bigg\}.
\end{aligned} \tag{3.45}
$$

For the normalization constant we use the value

$$
\mathcal{N}\left[\det\left(\frac{\Xi}{\hbar}\right)\right]^{-\frac{1}{2}} = \left(\frac{M\omega}{2\pi i\hbar\sin\omega T}\right)^{\frac{1}{2}}, \tag{3.46}
$$

arising from the normalization condition in Eq.(2.40). One should keep in mind that the normalization condition (2.40) holds true only after setting the ground state energy eigenvalue to zero. This is achieved through the global shift of the energy eigenvalue spectrum $E_n \to E_n - 1/2\hbar\omega$, $n = 0, 1, \ldots$. As a consequence, $\exp\left(\frac{i}{\hbar}E_0 T\right) = 1$. This information has been fed into the right hand side of Eq.(3.35) when looking for $\mathcal{N}$. One can check that (3.46) provides the right value for $\mathcal{N}$. Indeed, by substituting Eq.(3.46) into Eq.(3.16) we find

$$
{}_H\langle q_f, t_f | q_i, t_i\rangle_H^{J=0} = \left(\frac{M\omega}{2\pi i\hbar\sin\omega T}\right)^{\frac{1}{2}} e^{\frac{i}{\hbar}S^{J=0}[q_{cl}]}, \tag{3.47}
$$

[3]We invoked, whenever necessary, the identity

$$
\begin{aligned}
&\int_{t_i}^{t_f} dt \int_{t_i}^{t_f} dt' \sin\omega(t_f-t)\sin\omega(t'-t_i) \\
&\equiv \int_{t_i}^{t_f} dt \int_{t_i}^{t_f} dt' \left[\theta(t-t') + \theta(t'-t)\right] \sin\omega(t_f-t)\sin\omega(t'-t_i) \\
&= \int_{t_i}^{t_f} dt \int_{t_i}^{t_f} dt'\theta(t-t')\left[\sin\omega(t_f-t)\sin\omega(t'-t_i)\right. \\
&+ \left.\sin\omega(t_f-t')\sin\omega(t-t_i)\right].
\end{aligned}
$$

in agreement with the result obtained in Chapter 1 (see Eq.(1.184)). As a by-product, we have corroborated that different definitions of the phase space path integral lead, as they must, to the same result.

By means of trigonometric identities we can confirm that the real and imaginary parts of the exponent in Eq.(3.45) are, respectively, given by

$$-\frac{1}{2M\omega\hbar}\int_{t_i}^{t_f} dt \int_{t_i}^{t_f} dt'\theta(t-t')J(t)\cos\omega(t-t')J(t'), \tag{3.48}$$

and

$$\frac{1}{2M\omega\hbar}\int_{t_i}^{t_f} dt \int_{t_i}^{t_f} dt'\theta(t-t')J(t)\sin\omega(t-t')J(t'). \tag{3.49}$$

In turn, this allows us to cast the right hand side of Eq.(3.45) as in [Feynman (1950)], namely,

$$\begin{aligned}\tilde{\mathcal{U}}_0[J|t_f,t_i] &= \exp\left\{-\frac{1}{2M\omega\hbar}\right.\\ &\times \left.\int_{t_i}^{t_f} dt \int_{t_i}^{t_f} dt'\theta(t-t')J(t)\exp\left[-i\omega(t-t')\right]J(t')\right\}.\end{aligned} \tag{3.50}$$

It is of interest to introduce now the function $\Delta_F(t,t')$,

$$\begin{aligned}\Delta_F(t,t') &\equiv \frac{1}{2iM\omega}\left[\theta(t-t')\,e^{-i\omega(t-t')} + \theta(t'-t)\,e^{-i\omega(t'-t)}\right]\\ &= \frac{1}{2iM\omega}\,e^{-i\omega|t-t'|},\end{aligned} \tag{3.51}$$

in terms of which Eq.(3.50) may be rewritten as

$$\tilde{\mathcal{U}}_0[J|t_f,t_i] = \exp\left[-\frac{i}{2\hbar}\int_{t_i}^{t_f} dt \int_{t_i}^{t_f} dt' J(t)\Delta_F(t,t')J(t')\right]. \tag{3.52}$$

Notice that the integrands in Eqs.(3.50) and (3.52) *do not depend* on t_f or t_i. Hence, the ground state persistence amplitude when the system evolves in time from $t_i = -\infty$ to $t_f = +\infty$ is given by

$$\begin{aligned}&\tilde{\mathcal{U}}_0[J|+\infty,-\infty]\\ &= \exp\left[-\frac{i}{2\hbar}\int_{-\infty}^{+\infty} dt \int_{-\infty}^{+\infty} dt' J(t)\Delta_F(t,t')J(t')\right].\end{aligned} \tag{3.53}$$

We close this section by showing that the ground state-ground state transition probability,

$$|\tilde{\mathcal{U}}_0[J|t_f,t_i]|^2 = \exp\left\{-\frac{1}{M\omega\hbar} \times \int_{t_i}^{t_f} dt \int_{t_i}^{t_f} dt' \theta(t-t') J(t) \cos\omega(t-t') J(t')\right\}, \tag{3.54}$$

verifies, as required, $0 \leq |\tilde{\mathcal{U}}_0[J|t_f,t_i]|^2 \leq 1$. Indeed, by taking advantage of the symmetry of the product $J(t)\cos\omega(t-t')J(t')$ under the exchange $t \leftrightarrow t'$ we can rewrite Eq.(3.54) as

$$|\tilde{\mathcal{U}}_0[J|t_f,t_i]|^2 = \exp\left\{-\frac{1}{2M\omega\hbar} \times \int_{t_i}^{t_f} dt \int_{t_i}^{t_f} dt' J(t) \cos\omega(t-t') J(t')\right\} \tag{3.55}$$

or, equivalently,

$$|\tilde{\mathcal{U}}_0[J|t_f,t_i]|^2 = \exp\left\{-\frac{1}{2M\omega\hbar} \times \left[\left(\int_{t_i}^{t_f} dt J(t) \cos\omega t\right)^2 + \left(\int_{t_i}^{t_f} dt J(t) \sin\omega t\right)^2\right]\right\}, \tag{3.56}$$

QED.

3.2 More on the computation of the generating functional for the one dimensional harmonic oscillator

The computation of $\tilde{\mathcal{U}}_0[J|t_f,t_i]$ demanded the knowledge of the ground state wave function. This information is provided by the operator formulation of quantum mechanics. We emphasize here that, nevertheless, $\tilde{\mathcal{U}}_0[J|+\infty,-\infty]$ can be computed through a systematics solely based on the functional formalism. We will be showing, afterwards, that this holds true not only for the harmonic oscillator but also in the general case. In this sense, the limit $t_f \to +\infty$, $t_i \to -\infty$ is special.

We shall first discuss about the link connecting the function $\Delta_F(t_1,t_2)$, defined at Eq.(3.51) and showing up in Eq.(3.53), with the Green function

$(\Delta_\Xi(t,t'))$ of the local operator in Eq.(3.15) defined over the infinite time interval, namely,

$$M\left(-\frac{d^2}{dt^2}-\omega^2\right)\Delta_\Xi(t,t') = \delta(t-t'), \quad -\infty < t,t' < +\infty. \tag{3.57}$$

Let us focus on solving Eq.(3.57). The situation is different from that faced in the case of the finite time interval $t_i < t, t' < t_f$ where the problem possesses the *unique solution* quoted in Eq.(3.27). Both linearity and invariance under time translations suggest the Fourier transform as the appropriate tool for solving Eq.(3.57). We shall denote by $\Delta_\Xi(p)$ the Fourier transform of $\Delta_\Xi(t-t')$. Thus

$$\Delta_\Xi(t,t') = \Delta_\Xi(t-t') = \frac{1}{2\pi}\int_{-\infty}^{+\infty} dp\,\Delta_\Xi(p)\,e^{ip(t-t')}. \tag{3.58}$$

By returning with Eq.(3.58) into Eq.(3.57) we get

$$\Delta_\Xi(p) = \frac{1}{M}\frac{1}{p^2-\omega^2}, \tag{3.59}$$

which, when substituted back into Eq.(3.58), yields

$$\Delta_\Xi(t-t') = \frac{1}{2\pi M}\int_{-\infty}^{+\infty} dp\,\frac{e^{ip(t-t')}}{p^2-\omega^2}. \tag{3.60}$$

The presence of poles on the contour of integration signalizes that the right hand side in Eq.(3.60) is a mathematically ill-defined expression. However, Eq.(3.60) becomes meaningful after shifting the poles away from the real p-axis. This can be done in different ways, all of which associated with a specific type of boundary conditions. Let us investigate the effect provoked by the following replacement

$$p^2-\omega^2 \longrightarrow p^2-\omega^2+i\epsilon. \tag{3.61}$$

Here, $\Re\epsilon > 0$, $\Im\epsilon = 0$ while the limit $\epsilon \to 0^+$ is to be taken at the end of the calculations. This will be referred to as the *Feynman prescription*. The position of the displaced poles,

$$p_{1,2} = \pm\sqrt{\omega^2-i\epsilon}, \tag{3.62}$$

in the p complex plane is shown in Fig.3.1. The *regularized* Green function

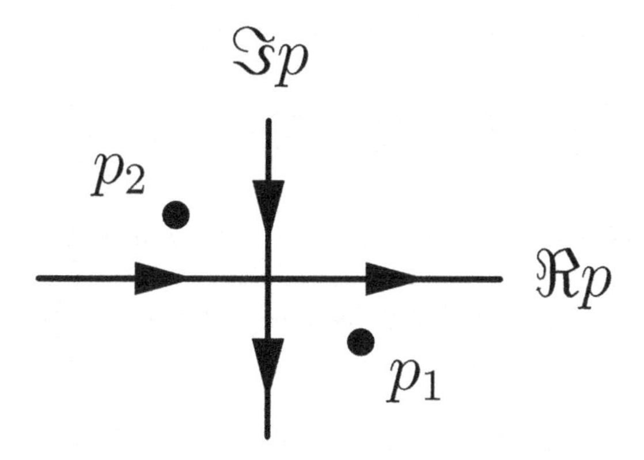

Fig. 3.1 Feynman prescription.

$$\Delta_{\Xi_\epsilon}(t-t') \equiv \frac{1}{2\pi M}\int_{-\infty}^{+\infty} dp\, \frac{e^{ip(t-t')}}{p^2-\omega^2+i\epsilon}, \tag{3.63}$$

is a mathematically well-defined entity. The factorization

$$p^2-\omega^2+i\epsilon = (p-p_1)(p-p_2), \tag{3.64}$$

allows for computing the integral in Eq.(3.63) via the residue theorem. For $t-t'>0$ the contour of integration may be closed in the upper half plane ($\Im p > 0$). In this case only the pole p_2 contributes and, therefore,

$$\lim_{\epsilon\to 0^+} \Delta_{\Xi_\epsilon}(t-t') = \frac{1}{2iM\omega}\, e^{-i\omega(t-t')}. \tag{3.65}$$

For $t-t'<0$ the contour of integration may be closed in the lower half plane ($\Im p < 0$) implying that only the pole p_1 contributes to the integral and

$$\lim_{\epsilon\to 0^+} \Delta_{\Xi_\epsilon}(t-t') = \frac{1}{2iM\omega}\, e^{+i\omega(t-t')}. \tag{3.66}$$

The results in Eqs.(3.65) and (3.66) can be combined into the single expression

$$\begin{aligned}\lim_{\epsilon\to 0^+} \Delta_{\Xi_\epsilon}(t-t') &= \frac{1}{2iM\omega} \\ &\times \left[\theta(t-t')e^{-i\omega(t-t')} + \theta(t'-t)e^{-i\omega(t'-t)}\right].\end{aligned} \tag{3.67}$$

Hence, from Eqs.(3.67) and (3.51) we obtain

$$\lim_{\epsilon \to 0^+} \Delta_{\Xi_\epsilon}(t - t') = \Delta_F(t - t') . \tag{3.68}$$

In words: the function Δ_F, emerging along the calculation of $\tilde{\mathcal{U}}_0[J| + \infty, -\infty]$, is the Green function of the operator (3.15) obeying the *boundary conditions imposed by the Feynman prescription.*

We switch now to the apparently disconnected task of computing the functional integral

$$\mathcal{Z}[J] \equiv \int [\mathcal{D}q] \exp\left\{\frac{i}{\hbar} \tilde{S}^J[q]\right\} , \tag{3.69}$$

with

$$\tilde{S}^J[q] \equiv \frac{1}{2} \int_{-\infty}^{+\infty} dt \int_{-\infty}^{+\infty} dt' \, q(t) \, \Xi(t,t') \, q(t') + \int_{-\infty}^{+\infty} dt \, q(t) \, J(t) , \tag{3.70}$$

over the collection of histories $\{q(t)\}$ obeying the boundary conditions deriving from the Feynman prescription. To that end, we start again by changing the integration variable as follows

$$q(t) = \tilde{q}^J_{cl}(t) + \eta(t) . \tag{3.71}$$

Here, $\tilde{q}^J_{cl}(t)$ solves the following classical equation

$$\left. \frac{\delta \tilde{S}^J[q]}{\delta q(t)} \right|_{q=\tilde{q}^J_{cl}} = 0 . \tag{3.72}$$

From Eqs.(3.70) and (3.72) and by setting the homogeneous part of the solution defined by Eq.(3.72) to zero we find

$$\tilde{q}^J_{cl}(t) = - \int_{-\infty}^{+\infty} dt \Delta_F(t,t') \, J(t') . \tag{3.73}$$

This solution obeys the boundary conditions deriving from the Feynman prescription because they are already *built in* $\Delta_F(t,t')$. At last, we enforce the just mentioned boundary conditions over all configurations $q(t)$ within the domain of integration by choosing

$$\eta(t = \pm\infty) = 0 . \tag{3.74}$$

By returning with Eq.(3.73) into Eq.(3.71) and, afterwards, into Eq.(3.69) we end up with

$$
\begin{aligned}
\mathcal{Z}[J] = & \int [\mathcal{D}\eta] \exp\left\{ \frac{i}{2\hbar} \int_{-\infty}^{+\infty} dt \int_{-\infty}^{+\infty} dt' \eta(t) \Xi(t,t') \eta(t') \right\} \\
& \times \exp\left\{ -\frac{i}{2\hbar} \int_{-\infty}^{+\infty} dt \int_{-\infty}^{+\infty} dt' J(t) \Delta_F(t,t') J(t') \right\} = \mathcal{N} \left[\det\left(\frac{\Xi}{\hbar} \right) \right]^{-\frac{1}{2}} \\
& \times \exp\left\{ -\frac{i}{2\hbar} \int_{-\infty}^{+\infty} dt \int_{-\infty}^{+\infty} dt' J(t) \Delta_F(t,t') \, J(t') \right\} ,
\end{aligned}
\tag{3.75}
$$

where we have assumed that the result in Eq.(B.78) remains valid for $a \to -\infty$ and $b \to +\infty$.

We pinpoint that Eq.(3.53) can be written as

$$
\tilde{\mathcal{U}}_0[J| + \infty, -\infty] = \frac{\mathcal{Z}[J]}{\mathcal{Z}[J=0]} . \tag{3.76}
$$

Thus, we have reached Eq.(3.53) without recourse to results provided by the operator formulation of quantum mechanics. We will verify that the same holds true for the general case. We shall return to this subject further in section 3.5.

3.3 The canonical algebra and the functional formalism

The formulation of the quantum dynamics within the operator framework demands: 1) the algebra of observables and 2) a representation of this algebra. On the other hand, within the functional scheme the quantum theory is defined by the set of connected Green functions. We illustrate below how these alternative formulations make contact.

Let A and B be observables. It is not difficult to relate the equal time commutator $[A(t), B(t)]$ with the chronological product $\mathcal{T}(A(t)B(t'))$. Indeed,

$$
[A(t), B(t)] = \mathcal{T}(A(t)B(t')) \Big|_{t \downarrow t'} - \mathcal{T}(A(t)B(t')) \Big|_{t \uparrow t'} . \tag{3.77}
$$

Thus, it is possible to learn about the matrix element

$$
\langle E_0, t_f | \, [A(t), B(t)] \, | E_0, t_i \rangle \tag{3.78}
$$

by studying the two-point Green functions $\langle E_0, t_f | \mathcal{T}\left(A(t)B(t')\right) | E_0, t_i\rangle$.

Of particular importance are the canonical quantization rules, namely, when A and B are either Q or P. Correspondingly, we shall be looking for the two-point Green functions

$$\begin{aligned}
\omega^{11}(t_f,t,t',t_i) &\equiv \langle E_0,t_f|\mathcal{T}\{Q(t)\,Q(t')\}\,|E_0,t_i)\rangle \\
&= \left(\frac{\hbar}{i}\right)^2 \left.\frac{\delta^2\tilde{\mathcal{U}}_0\left[s|t_f,t_i\right]}{\delta J(t)\,\delta J(t')}\right|_{J=K=0}, \qquad (3.79a)\\
\omega^{1}{}_{1}(t_f,t,t',t_i) &\equiv \langle E_0,t_f|\mathcal{T}\{Q(t)\,P(t')\}\,|E_0,t_i)\rangle \\
&= \left(\frac{\hbar}{i}\right)^2 \left.\frac{\delta^2\tilde{\mathcal{U}}_0\left[s|t_f,t_i\right]}{\delta J(t)\,\delta K(t')}\right|_{J=K=0}, \qquad (3.79b)\\
\omega_{11}(t_f,t,t',t_i) &\equiv \langle E_0,t_f|\mathcal{T}\{P(t)\,P(t')\}\,|E_0,t_i)\rangle \\
&= \left(\frac{\hbar}{i}\right)^2 \left.\frac{\delta^2\tilde{\mathcal{U}}_0\left[s|t_f,t_i\right]}{\delta K(t)\,\delta K(t')}\right|_{J=K=0}. \qquad (3.79c)
\end{aligned}$$

For this purpose, one needs to know the Green functions generating functional when the fictitious sources of coordinates and momenta are both turned on. Hence, Eq.(3.3) must be substituted by

$$\begin{aligned}
{}_H\langle q_f,t_f|q_i,t_i\rangle_H^{(s)} &= \mathcal{N}\int[\mathcal{D}q]\int[\mathcal{D}p] \\
&\times \exp\left\{\frac{i}{\hbar}\int_{t_i}^{t_f} dt\left[p(t)\dot{q}(t) - \frac{p^2}{2M} - \frac{1}{2}M\omega^2q^2\right]\right. \\
&\left. + \frac{i}{\hbar}\int_{t_i}^{t_f} dt\, q(t)J(t) + \frac{i}{\hbar}\int_{t_i}^{t_f} dt\, p(t)K(t)\right\}. \qquad (3.80)
\end{aligned}$$

The computation of the functional momentum integral goes as before and yields

$$\begin{aligned}
{}_H\langle q_f,t_f|q_i,t_i\rangle_H^{(s)} &= \mathcal{N}\exp\left[\frac{iM}{2\hbar}\int_{t_i}^{t_f} dt K^2(t)\right] \\
&\times \int[\mathcal{D}q]\exp\left\{\frac{i}{\hbar}S^{(s)}[q]\right\}, \qquad (3.81)
\end{aligned}$$

where

$$S^{(s)}[q] = \int_{t_i}^{t_f} dt\left(\frac{M}{2}\dot{q}^2 + M\dot{q}K - \frac{1}{2}M\omega^2q^2 + Jq\right). \qquad (3.82)$$

We shall assume that the fictitious source $K(t)$ vanishes outside the interval $[t_i, t_f]$. Then, after a by part integration $S^{(s)}[q]$ can be cast

$$S^{(s)}[q] = \int_{t_i}^{t_f} dt \left[\frac{M}{2}\dot{q}^2 - \frac{1}{2}M\omega^2 q^2 + q\left(J - M\dot{K}\right)\right] . \tag{3.83}$$

We look next for the corresponding generating functional of Green functions. To that end, we invoke again Eq.(2.68). We leave as an exercise for the reader to corroborate that

$$\tilde{\mathcal{U}}_0[s|t_f, t_i] = \exp\left[\frac{iM}{2\hbar}\int_{t_i}^{t_f} dt K^2(t)\right] \exp\left\{ -\frac{i}{2\hbar}\int_{t_i}^{t_f} d\tau \int_{t_i}^{t_f} d\tau' \right.$$

$$\left. \times \left[J(\tau) - M\dot{K}(\tau)\right] \Delta_F(\tau, \tau') \left[J(\tau') - M\dot{K}(\tau')\right] \right\} . \tag{3.84}$$

The calculation of the functional derivatives in Eqs.(3.79) yields

$$\left(\frac{\hbar}{i}\right)^2 \left.\frac{\delta^2 \tilde{\mathcal{U}}_0[s|t_f, t_i]}{\delta J(t)\, \delta J(t')}\right|_{s=0} = i\hbar \Delta_F(t, t') , \tag{3.85a}$$

$$\left(\frac{\hbar}{i}\right)^2 \left.\frac{\delta^2 \tilde{\mathcal{U}}_0[s|t_f, t_i]}{\delta J(t)\, \delta K(t')}\right|_{s=0} = i\hbar M \frac{\partial \Delta_F(t, t')}{\partial t'} , \tag{3.85b}$$

$$\left(\frac{\hbar}{i}\right)^2 \left.\frac{\delta^2 \tilde{\mathcal{U}}_0[s|t_f, t_i]}{\delta K(t) \delta K(t')}\right|_{s=0} = i\hbar M \left[M\frac{\partial^2 \Delta_F(t, t')}{\partial t \partial t'} - \delta(t - t')\right] . \tag{3.85c}$$

In accordance with Eq.(3.77) the equal time commutator is contributed by the discontinuity of the corresponding chronological product at $t = t'$. From Eq.(3.51) it follows that $\Delta_F(t - t')$ is continuous at $t = t'$, i.e.,

$$\left.\Delta_F(t, t')\right|_{t\downarrow t'} = \left.\Delta_F(t, t')\right|_{t\uparrow t'} . \tag{3.86}$$

Hence, in accordance with Eqs.(3.85a) and (3.79a), the same holds true for the matrix element

$$\langle E_0, t_f | \mathcal{T}\left\{Q(t)\, Q(t')\right\} | E_0, t_i \rangle$$

which along with Eq.(3.77) lead us to

$$\langle E_0, t_f | \left[Q(t), Q(t)\right] | E_0, t_i \rangle = 0 . \tag{3.87}$$

This is consistent with $[Q(t), Q(t)] = 0$.

Meanwhile, by starting again from Eq.(3.51) we find

$$\frac{\partial \Delta_F(t,t')}{\partial t'} = \frac{1}{2M}\left[\theta(t-t')e^{-i\omega(t-t')} - \theta(t'-t)e^{-i\omega(t'-t)}\right], \quad (3.88)$$

which is not continuous at $t = t'$. In fact,

$$i\hbar M\left[\left.\frac{\partial \Delta_F(t,t')}{\partial t'}\right|_{t\downarrow t'} - \left.\frac{\partial \Delta_F(t,t')}{\partial t'}\right|_{t\uparrow t'}\right] = i\hbar \neq 0\,. \quad (3.89)$$

Then, from Eqs.(3.85b), (3.79b) and (3.77) it follows that

$$\langle E_0, t_f|\,[Q(t), P(t)]\,|E_0, t_i\rangle \;=\; i\hbar\,, \quad (3.90)$$

in agreement with the canonical commutation rule $[Q(t), P(t)] = i\hbar I$.

At last, one can corroborate that

$$M\frac{\partial^2 \Delta_F(t,t')}{\partial t \partial t'} - \delta(t-t') \quad (3.91)$$

is continuous at $t = t'$, which on account of Eqs.(3.85c), (3.79c) and (3.77) yields

$$\langle E_0, t_f|\,[P(t), P(t)]\,|E_0, t_i\rangle \;=\; 0\,, \quad (3.92)$$

as required by $[P(t), P(t)] = 0$.

The operator and the functional formulations of quantum mechanics make contact in an indirect way; discontinuities in the Green functions signalize for the presence of non-commuting operators.

3.4 The one dimensional free particle

We address now the problem of computing the ground state persistence amplitude for systems possessing only a continuous energy spectrum. The prototype of such kind of systems is the one dimensional free particle which will be discussed here.

The computation referred above suffers from the drawback that the ground state is of infinite norm. To bypass this difficulty we shall instead be looking for the transition amplitude between finite norm initial and final states

$$|\phi(t_i)\rangle \to |\psi(t_f)\rangle \, .$$

Correspondingly, Eq.(2.68) is replaced by [4]

$$\tilde{\mathcal{U}}[J|t_f, t_i] = \frac{1}{2\pi\hbar} \int_{-\infty}^{+\infty} dp_f \int_{-\infty}^{+\infty} dp_i \int_{-\infty}^{+\infty} dq_f \int_{-\infty}^{+\infty} dq_i$$

$$\psi^{\star}(p_f)\, e^{-\frac{i}{\hbar} p_f q_f}\; {}_H\langle q_f, t_f | q_i, t_i\rangle_H^J \; e^{+\frac{i}{\hbar} p_i q_i}\, \phi(p_i) \, , \tag{3.93}$$

where $\psi(p_f) \equiv \; _S\langle p_f|\psi\rangle_S$, $\phi(p_i) \equiv \; _S\langle p_i|\phi_i\rangle_S$,

$${}_H\langle q_f, t_f | q_i, t_i\rangle_H^J = \mathcal{N} \int \left[\frac{\mathcal{D}q}{\sqrt{2\pi i\hbar}} \right] e^{\frac{i}{\hbar} S^J[q]} \, , \tag{3.94}$$

and

$$S^J[q] = \int_{t_i}^{t_f} dt \left[\frac{1}{2} M \dot{q}^2(t) + q(t) J(t) \right] . \tag{3.95}$$

The computation of the functional integral in Eq.(3.94) goes through as in the case of the harmonic oscillator. The result is

$${}_H\langle q_f, t_f | q_i, t_i\rangle_H^J = \mathcal{N} \left[\det \left(\frac{\Xi}{\hbar} \right) \right]^{-\frac{1}{2}} e^{\frac{i}{\hbar} S^J[q_{cl}]} \, , \tag{3.96}$$

where Ξ denotes here the local operator

$$\Xi(t, t') \equiv -M \frac{d^2}{dt^2} \delta(t - t') \, . \tag{3.97}$$

Also, the configuration q_{cl} solves the classical equation of motion arising from (3.95), namely,

$$M \ddot{q}_{cl} = J \, , \tag{3.98}$$

under the boundary conditions

$$q_{cl}(t_i) = q_i \, , \qquad q_{cl}(t_f) = q_f \, . \tag{3.99}$$

[4]It is important to keep in mind that, as far as classical mechanics is concerned, the system under analysis is just a particle of mass M moving along a straight line under the action of the time dependent external force $J(t)$.

We can check that

$$q_{cl}(t) = \frac{1}{T}\left[(q_f - q_i)\,t + q_i t_f - q_f t_i\right] + \frac{1}{M}\int_{t_i}^{t_f} dt' D_F(t,t')J(t')\,, \qquad (3.100)$$

where

$$D_F(t,t') = -\frac{1}{T}\left[\theta(t-t')(t-t_i)(t_f-t') + \theta(t'-t)(t'-t_i)(t_f-t)\right] \qquad (3.101)$$

denotes the Green function of the operator d^2/dt^2 fulfilling

$$D_F(t_f,t') = 0\,, \qquad D_F(t_i,t') = 0\,. \qquad (3.102)$$

We emphasize that $D_F(t,t')$ is *unique*. Hence, the eigenvalue problem of the operator d^2/dt^2, whose eigenfunctions vanish at the end points of the interval $[t_i, t_f]$, is not afflicted by *zero modes*. This secures the existence of

$$\left[\det\left(\frac{\Xi}{\hbar}\right)\right]^{-\frac{1}{2}}\,.$$

The evaluation of $S^J[q_{cl}]$ renders [5]

$$S^J[q_{cl}] = \frac{M}{2T}\left[q_f^2 + q_i^2 - 2q_f q_i + \frac{2q_f}{M}\int_{t_i}^{t_f} dt(t-t_i)J(t) + \frac{2q_i}{M}\int_{t_i}^{t_f} dt(t_f-t)J(t) - \frac{2}{M^2}\int_{t_i}^{t_f} dt\int_{t_i}^{t_f} dt' J(t)\,\theta(t-t')(t_f-t)(t'-t_i)J(t')\right]\,. \qquad (3.103)$$

By going back with this last expression into Eq.(3.96) and with the result thus obtained into Eq.(3.93) we have that

$$\tilde{\mathcal{U}}[J|t_f,t_i] = \frac{1}{2\pi\hbar}\mathcal{N}\left[\det\left(\frac{\Xi}{\hbar}\right)\right]^{-\frac{1}{2}} \times \exp\left[-\frac{i}{\hbar MT}\int_{t_i}^{t_f} dt\int_{t_i}^{t_f} dt' J(t)\theta(t-t')(t_f-t)(t'-t_i)J(t')\right] \times \int_{-\infty}^{+\infty} dp_f\psi^\star(p_f)\int_{-\infty}^{+\infty} dp_i\phi(p_i)\,\Lambda[J|p_f,p_i;t_f,t_i]\,, \qquad (3.104)$$

[5] Notice that Eq.(3.103) follows by taking the limit $\omega \to 0$ in Eq.(3.34).

where

$$\Lambda[J|p_f, p_i; t_f, t_i] \equiv \int_{-\infty}^{+\infty} dq_f \int_{-\infty}^{+\infty} dq_i$$
$$\times \exp\left(-aq_f^2 - aq_i^2 + bq_f q_i + cq_f + dq_i\right) \tag{3.105}$$

and

$$a \equiv -\frac{iM}{2\hbar T}, \tag{3.106a}$$
$$b \equiv -\frac{iM}{\hbar T}, \tag{3.106b}$$
$$c \equiv \frac{i}{\hbar}\left[-p_f + \frac{1}{T}\int_{t_i}^{t_f} dt(t - t_i)\, J(t)\right], \tag{3.106c}$$
$$d \equiv \frac{i}{\hbar}\left[p_i + \frac{1}{T}\int_{t_i}^{t_f} dt(t_f - t)\, J(t)\right]. \tag{3.106d}$$

As in the case of the harmonic oscillator, the computation of the integrals in Eq.(3.105) is facilitated by the introduction of new integration variables already defined in Eq.(3.38). We, then, find

$$\Lambda[J|p_f, p_i; t_f, t_i] = \Lambda_1[J|p_f, p_i; t_f, t_i]\, \Lambda_2[J|p_f, p_i; t_f, t_i], \tag{3.107}$$

where

$$\Lambda_1 = \int_{-\infty}^{+\infty} dx \exp\left[\left(-a + \frac{b}{2}\right) x^2 + \frac{1}{\sqrt{2}}(c + d)\, x\right], \tag{3.108a}$$
$$\Lambda_2 = \int_{-\infty}^{+\infty} dy \exp\left[\left(-a - \frac{b}{2}\right) y^2 + \frac{1}{\sqrt{2}}(d - c)\, y\right]. \tag{3.108b}$$

Thus, after taking into account Eq.(3.106) and the results in Appendix A we obtain

$$\Lambda_1 = 2\pi\hbar\sqrt{2}\,\delta\,(C + D)\,, \tag{3.109}$$

$$\Lambda_2 = \sqrt{\frac{i\pi\hbar T}{M}} \exp\left[-\frac{iT}{8\hbar M}(D - C)^2\right], \tag{3.110}$$

where we have introduced the definitions

$$C \equiv -p_f + \frac{1}{T}\int_{t_i}^{t_f} dt(t-t_i)\,J(t)\,, \tag{3.111a}$$

$$D \equiv p_i + \frac{1}{T}\int_{t_i}^{t_f} dt(t_f-t)\,J(t)\,. \tag{3.111b}$$

We return next with Eqs.(3.110), (3.109) into (3.107) which is, in turn, substituted back into Eq.(3.104). The normalization constant $\mathcal{N}$ is determined by demanding [6]

$$\tilde{\mathcal{U}}[J=0|t_f,t_i] = \int_{-\infty}^{+\infty} dp_f \int_{-\infty}^{+\infty} dp_i$$
$$\times\psi^\star(p_f)\exp\left(-\frac{i}{\hbar}\frac{p_f^2}{2M}T\right)\phi(p_i)\delta(-p_f+p_i)\,. \tag{3.112}$$

After some algebraic rearrangements we get

$$\tilde{\mathcal{U}}[J|t_f,t_i]$$
$$= \exp\left[\frac{i}{2\hbar M}\int_{t_i}^{t_f} d\tau\int_{t_i}^{t_f} d\tau' J(\tau)\theta(\tau-\tau')(\tau-\tau')J(\tau')\right]$$
$$\times\int_{-\infty}^{+\infty} dp_f\psi^\star(p_f)\int_{-\infty}^{+\infty} dp_i\phi(p_i)\exp\left\{-\frac{iT}{2\hbar M}p_f p_i\right.$$
$$\left.-\frac{i}{2\hbar M}\left[(p_f t_f + p_i t_i)\int_{t_i}^{t_f} d\tau J(\tau) - (p_f+p_i)\int_{t_i}^{t_f} d\tau\tau J(\tau)\right]\right\}$$
$$\delta\left(p_f - p_i - \int_{t_i}^{t_f} d\tau J(\tau)\right)\,. \tag{3.113}$$

The presence of the generalized Dirac delta function, on the right hand side of Eq.(3.113), secures that the change in the linear momentum equals the impulse transferred from the fictitious force $J(t)$ to the particle during

---

[6] We leave it as an exercise for the reader to verify that

$$\mathcal{N}\left[\det\left(\frac{\Xi}{\hbar}\right)\right]^{-\frac{1}{2}} = \sqrt{\frac{M}{2\pi i\hbar T}}\,.$$

the time interval T. We can see that

$$\begin{aligned}
&\tilde{\mathcal{U}}[J|t_f, t_i] \\
&= \exp\left[\frac{i}{\hbar M}\int_{t_i}^{t_f} d\tau \int_{t_i}^{t_f} d\tau' J(\tau)\theta(\tau-\tau')(t_i-\tau')J(\tau')\right] \\
&\times \int_{-\infty}^{+\infty} dp_f \psi^\star(p_f)\phi\left[p_f - \int_{t_i}^{t_f} d\tau J(\tau)\right] \\
&\times \exp\left[-\frac{iT}{2\hbar M}p_f^2 - \frac{i}{\hbar M}p_f \int_{t_i}^{t_f} d\tau\,(t_i-\tau)\,J(\tau)\right]. \qquad (3.114)
\end{aligned}$$

In particular, the amplitude for the particle to be in the linear momentum eigenstate $\psi(p_f) = \delta(p_f - p)$ at the time $t = t_f$ is

$$\begin{aligned}
&\tilde{\mathcal{U}}_p[J|t_f, t_i] \\
&= \exp\left[\frac{i}{\hbar M}\int_{t_i}^{t_f} d\tau \int_{t_i}^{t_f} d\tau' J(\tau)\theta(\tau-\tau')(t_i-\tau')J(\tau')\right] \\
&\times \phi\left[p - \int_{t_i}^{t_f} d\tau J(\tau)\right] \\
&\times \exp\left[-\frac{iT}{2\hbar M}p^2 - \frac{i}{\hbar M}p \int_{t_i}^{t_f} d\tau\,(t_i-\tau)\,J(\tau)\right], \qquad (3.115)
\end{aligned}$$

while the probability of the particle be detected in the final state with linear momentum in the interval between p and $p + dp$ is

$$|\tilde{\mathcal{U}}_p[J|t_f, t_i]|^2 dp = \left|\phi\left[p - \int_{t_i}^{t_f} d\tau J(\tau)\right]\right|^2 dp\,. \qquad (3.116)$$

But this is, precisely, the probability for finding the particle in the initial state with linear momentum in the interval between $p - \int_{t_i}^{t_f} d\tau J(\tau)$ and $p - \int_{t_i}^{t_f} d\tau J(\tau) + dp$. This shift in momentum summarizes the effect provoked by the fictitious force on the free particle.

3.5 The general case

The question now is whether or not the results derived in connection with simple systems above are of general validity.

We start again by addressing the problem of computing the Green functions generating functional for a finite interval $[t_f, t_i]$. The strategy will be similar to that in [Abers and Lee (1973)]. It is based on introducing the intermediate instants of time t'_i and t'_f such that

$$t_i < t'_i < t'_f < t_f \,. \tag{3.117}$$

We choose the fictitious source K to vanish at all times, while the fictitious source J vanishes outside the interval $[t'_i, t'_f]$. It will prove convenient to write

$$\begin{aligned} {}_H\langle q_f, t_f | q_i, t_i \rangle_H^J &= \int_{-\infty}^{+\infty} dq'_f \int_{-\infty}^{+\infty} dq'_i \\ & {}_H\langle q_f, t_f | q'_f, t'_f \rangle_H \; {}_H\langle q'_f, t'_f | q'_i, t'_i \rangle_H^J \; {}_H\langle q'_i, t'_i | q_i, t_i \rangle_H \,. \end{aligned} \tag{3.118}$$

The absence of the superscript J signalizes that the corresponding transition ($t_i \to t'_i$ and $t'_f \to t_f$) is not being acted upon by the fictitious source J. Hence, we can write (see the developments in section 1.1)

$$\begin{aligned} {}_H\langle q_f, t_f | q'_f, t'_f \rangle_H &= {}_S\langle q_f | e^{-\frac{i}{\hbar} H_S (t_f - t'_f)} | q'_f \rangle_S \\ &= \sum_{n=0}^{\infty} e^{-\frac{i}{\hbar} E_n (t_f - t'_f)} \, \psi_n(q_f) \, \psi_n^*(q'_f) = e^{-\frac{i}{\hbar} E_0 (t_f - t'_f)} \, \psi_0(q_f) \, \psi_0^*(q'_f) \\ &+ \sum_{n=1}^{\infty} e^{-\frac{i}{\hbar} E_n (t_f - t'_f)} \, \psi_n(q_f) \, \psi_n^*(q'_f) \,, \end{aligned} \tag{3.119}$$

where a complete set of energy eigenstates was introduced. Here, $\psi_n(q)$ is the nth energy eigenfunction and E_n the corresponding eigenvalue. Furthermore, $E_n > E_0$, where E_0 denotes the ground state energy eigenvalue [7]. The global displacement of the energy spectrum $E_n \to E_n - E_0, \forall n$, already referred to in Chapter 2, sets $E_0 = 0$ and enables one to rewrite Eq.(3.119) as

$$\begin{aligned} {}_H\langle q_f, t_f | q'_f, t'_f \rangle_H &= \psi_0(q_f) \, \psi_0^*(q'_f) \\ &+ \sum_{n=1}^{\infty} e^{-\frac{i}{\hbar} E_n (t_f - t'_f)} \, \psi_n(q_f) \, \psi_n^*(q'_f) \,, \end{aligned} \tag{3.120}$$

where $E_n > 0$.

We shall next take the limit $t_f \to +\infty$ on both sides of Eq.(3.120). This is not a well-defined mathematical operation due to the presence of

[7] The existence of this energy gap will prove essential for further developments in this section.

oscillatory factors. The recourse to a regularization scheme becomes unavoidable. In [Abers and Lee (1973)] and [Ramond (1990)] it appears based on the continuation of T along the positive imaginary axis, while here we appeal to the *analytic extension* of Eq.(3.120) to the complex energy plane, namely,

$$E_n \to E_n - i\,\epsilon\,, \quad \forall\, n\,, \qquad \Im\epsilon = 0\,, \qquad \Re\epsilon > 0\,. \tag{3.121}$$

The limit $\epsilon \to 0^+$ being taken at the end of the calculations. Hence,

$$\lim_{t_f \to +\infty} {}_H\langle q_f, t_f | q'_f, t'_f \rangle_H \;=\; \psi_0(q_f)\,\psi_0^*(q'_f)\,. \tag{3.122}$$

Also

$$\lim_{t_i \to -\infty} {}_H\langle q'_i, t'_i | q_i, t_i \rangle_H \;=\; \psi_0^*(q_i)\,\psi_0(q'_i)\,. \tag{3.123}$$

From Eqs.(3.118) and (2.68) we, then, find

$$\begin{aligned}
\lim_{\substack{t_f \to +\infty \\ t_i \to -\infty}} {}_H\langle q_f, t_f | q_i, t_i \rangle_H^J \;&=\; \psi_0(q_f)\,\psi_0^*(q_i) \\
&\times \int_{-\infty}^{+\infty} dq'_f \int_{-\infty}^{+\infty} dq'_i \psi_0^*(q'_f)\,{}_H\langle q'_f, t'_f | q'_i, t'_i \rangle_H^J\,\psi_0(q'_i) \\
&= \psi_0(q_f)\,\psi_0^*(q_i)\,\tilde{\mathcal{U}}_0\left[J | t'_f, t'_i\right]\,.
\end{aligned} \tag{3.124}$$

Moreover, by taking into account Eq.(2.82), specialized for the situation under analysis, we arrive at

$$\begin{aligned}
&\mathcal{N} \int [\mathcal{D}q] \int [\mathcal{D}p] \\
&\times \exp\left\{\frac{i}{\hbar}\int_{-\infty}^{+\infty} dt\,\left[p_j(t)\dot{q}^j(t) \;-\; h(q(t),p(t)) \;+\; q^j(t)J_j(t)\right]\right\} \\
&= \psi_0(q_f)\,\psi_0^*(q_i)\,\tilde{\mathcal{U}}_0\left[J | t'_f, t'_i\right]\,.
\end{aligned} \tag{3.125}$$

The inequality (3.117) does not prevent us from letting t'_f and $-t'_i$ tend to infinity since we have already set $t_f \;=\; +\infty$ and $t_i \;=\; -\infty$. Therefore, Eq.(3.125) gives rise to

$$\begin{aligned}
&\mathcal{N} \int [\mathcal{D}q] \int [\mathcal{D}p] \\
&\times \exp\left\{\frac{i}{\hbar}\int_{-\infty}^{+\infty} dt\,\left[p_j(t)\dot{q}^j(t) \;-\; h(q(t),p(t)) \;+\; q^j(t)J_j(t)\right]\right\} \\
&= \psi_0(q_f)\,\psi_0^*(q_i)\,\tilde{\mathcal{U}}_0\left[J | +\infty, -\infty\right]\,.
\end{aligned} \tag{3.126}$$

When the fictitious source acts for a finite time interval, one is to include such restriction within the mathematical structure of the function $J(t)$.

Before proceeding, some clarifications are in order. We shall restrict ourselves to deal with systems whose Hamiltonian is of the following form

$$h(q,p) = \frac{p_j\, p_j}{2M} + V(q)\,, \tag{3.127}$$

where $V(q) = V(q^1, \ldots, q^N)$ is an analytic function of the coordinates. Under this assumption the functional integrals on p and q decouple. Indeed,

$$p_j(t)\dot{q}^j(t) - \frac{p_j\, p_j}{2M} = -\frac{1}{2M}\left(p_j - M\dot{q}^j\right)\left(p_j - M\dot{q}^j\right) + \frac{M}{2}\,\dot{q}^j\dot{q}^j\,, \tag{3.128}$$

which enables us to find

$$\mathcal{N}\int[\mathcal{D}q]\int[\mathcal{D}p] \times \exp\left\{\frac{i}{\hbar}\int_{-\infty}^{+\infty} dt\,\left[p_j(t)\dot{q}^j(t) - h(q(t),p(t)) + q^j(t)J_j(t)\right]\right\} = \mathcal{N}\int[\mathcal{D}q]\exp\left\{\frac{i}{\hbar}S^J[q]\right\}, \tag{3.129}$$

where

$$S^J[q] = \int_{-\infty}^{+\infty} dt\, L(q(t),\dot{q}(t)) + \int_{-\infty}^{+\infty} dt\, q^j(t)J_j(t) \tag{3.130}$$

is the particle action and

$$L(q,\dot{q}) = \frac{1}{2}M\,\dot{q}^j\,\dot{q}^j - V(q) \tag{3.131}$$

the corresponding Lagrangian. Hence, Eq.(3.125) becomes

$$\mathcal{N}\int[\mathcal{D}q]\exp\left\{\frac{i}{\hbar}S^J[q]\right\} = \psi_0(q_f)\,\psi_0^*(q_i)\,\tilde{\mathcal{U}}_0\left[J\right| +\infty, -\infty\right]. \tag{3.132}$$

We must still determine the normalization constant $\mathcal{N}$. To that end, we set $J = 0$ and $\tilde{\mathcal{U}}_0\left[J = 0\right| +\infty, -\infty\right] = 1$. This yields

$$\mathcal{N} = \frac{\psi_0(q_f)\,\psi_0^*(q_i)}{\int[\mathcal{D}q]\exp\left\{\frac{i}{\hbar}S^{J=0}[q]\right\}} \tag{3.133}$$

which, when substituted into Eq.(3.132), leads to

$$\tilde{\mathcal{U}}_0\left[J|+\infty,-\infty\right] = \frac{\mathbf{Z[J]}}{\mathbf{Z[J=0]}}, \tag{3.134}$$

with

$$\mathbf{Z[J]} \equiv \int [\mathcal{D}q] \exp\left\{\frac{i}{\hbar} S^J[q]\right\}. \tag{3.135}$$

Needless to say, Eq.(3.76) is just a particular case of Eq.(3.134).

When the fictitious sources of momenta are turned on, Eq.(3.134) generalizes as follows

$$\tilde{\mathcal{U}}_0\left[J,K|+\infty,-\infty\right] = \frac{\mathbf{Z[J\,,K]}}{\mathbf{Z[J=0\,,K=0]}}, \tag{3.136}$$

where

$$\begin{aligned}\mathbf{Z[J\,,K]} &\equiv \int [\mathcal{D}q] \int [\mathcal{D}p] \\ &\times \exp\left\{\frac{i}{\hbar}\int_{-\infty}^{+\infty} dt \left[p_j(t)\dot{q}^j(t) \; - \; h(q(t),p(t))\right.\right. \\ &\left.\left. + q^j(t)J_j(t) \; + \; p_j K^j\right]\right\}. \end{aligned} \tag{3.137}$$

To conclude, $\tilde{\mathcal{U}}_0\left[J,K|+\infty,-\infty\right]$ can be found without recourse to information furnished by the operator formulation of quantum mechanics. This statement is of general validity.

3.6 Approximation methods

On general grounds, the functional integrals in Eq.(3.134) cannot be exactly evaluated. Approximate results can be obtained by means of a perturbative scheme known as either *steepest descent* or *stationary phase approximation.*

It is based on the fact that the classical trajectory $q^J_{cl}(t)$ makes the dominant contribution to the path integral. We then proceed towards the evaluation of the path integral, in the right hand side of Eq.(3.135), by resorting to the Taylor expansion of the functional $S^J[q]$ around $q = q^J_{cl}$. It reads [8]

$$S^J[q] = S^J[q_{cl}^J] + \int_{-\infty}^{+\infty} dt \left.\frac{\delta S^J[q]}{\delta q^k(t)}\right|_{q=q_{cl}^J} \eta^k(t)$$
$$+ \frac{1}{2!}\int_{-\infty}^{+\infty} dt_1 \int_{-\infty}^{+\infty} dt_2 \left.\frac{\delta^2 S^J[q]}{\delta q^{k_1}(t_1)\delta q^{k_2}(t_2)}\right|_{q=q_{cl}^J} \eta^{k_1}(t_1)\eta^{k_2}(t_2)$$
$$+ \sum_{n=3}^{\infty}\frac{1}{n!}\int_{-\infty}^{+\infty} dt_1 \ldots \int_{-\infty}^{\infty} dt_n \left.\frac{\delta^n S^J[q]}{\delta q^{k_1}(t_1)\ldots\delta q^{k_n}(t_n)}\right|_{q=q_{cl}^J}$$
$$\times \eta^{k_1}(t_1)\ldots\eta^{k_n}(t_n). \tag{3.138}$$

Here, $q_{cl}^J(t)$ solves the classical equations of motion

$$\left.\frac{\delta S^J[q]}{\delta q^k(t)}\right|_{q=q_{cl}^J} = \left.\frac{\delta S[q]}{\delta q^k(t)}\right|_{q=q_{cl}^J} + J_k = 0, \tag{3.139}$$

which, on account of Eqs.(3.130) and (3.131), is equivalent to

$$-M\frac{d^2 q_{cl}^J}{dt^2} - \left.\frac{\partial V(q)}{\partial q^j}\right|_{q=q_{cl}^J} + J(t) = 0. \tag{3.140}$$

Furthermore, the *quantum supplement* $\eta(t)$,

$$\eta(t) \equiv q(t) - q_{cl}^J(t), \tag{3.141}$$

is required to obey the boundary conditions

$$\eta(\pm\infty) = 0. \tag{3.142}$$

As in the case of the one dimensional harmonic oscillator (see section 3.2), the boundary conditions defining the domain of integration $\{q(t)\}$ are introduced by setting to zero the homogeneous part of the solution of the classical equation of motion (3.140), i.e.,

$$q_{cl}^{J=0}(t) = 0. \tag{3.143}$$

The behavior of $q_{cl}^J(t)$ at $t = \pm\infty$ is that of $q(t)$, since in accordance with Eqs.(3.141) and (3.142) we have that

$$q(t = \pm\infty) = q_{cl}^J(t = \pm\infty). \tag{3.144}$$

[8]We remind the reader about the summation convention over repeated indices.

We concentrate now on analyzing Eq.(3.138). In view of Eq.(3.139) the term linear in η drops out. Moreover, it will prove helpful to introduce the definitions

$$\begin{aligned}\Gamma_{k_1k_2}\left(q^J_{cl};t_1,t_2\right) &\equiv \left.\frac{\delta^2 S^J[q]}{\delta q^{k_1}(t_1)\delta q^{k_2}(t_2)}\right|_{q=q^J_{cl}} \\ &= \left[-M\,\delta_{k_1k_2}\frac{d^2}{dt_1^2} - V^{(2)}_{k_1k_2}(q^J_{cl}(t_1))\right]\delta(t_1-t_2)\end{aligned} \tag{3.145}$$

and

$$\begin{aligned}&\left.\frac{\delta^n S^J[q]}{\delta q^{k_1}(t_1)\ldots\delta q^{k_n}(t_n)}\right|_{q=q^J_{cl}} \\ &= -V^{(n)}_{k_1\ldots k_n}(q^J_{cl}(t_1))\,\delta(t_1-t_2)\ldots\delta(t_1-t_n)\,,\end{aligned} \tag{3.146}$$

where

$$V^{(n)}_{k_1\ldots k_n}(q^J_{cl}(t_1)) \equiv \left.\frac{\partial^n V(q(t_1))}{\partial q^{k_1}(t_1)\ldots\partial q^{k_n}(t_1)}\right|_{q=q^J_{cl}}. \tag{3.147}$$

By returning with Eqs.(3.145) and (3.146) into Eq.(3.138) we obtain

$$\begin{aligned}S^J[q] &= S^J[q^J_{cl}] \\ &+ \frac{1}{2!}\int_{-\infty}^{+\infty} dt_1 \int_{-\infty}^{+\infty} dt_2\,\eta^{k_1}(t_1)\,\Gamma_{k_1k_2}\left(q^J_{cl};t_1,t_2\right)\,\eta^{k_2}(t_2) \\ &- \sum_{n=3}^{\infty}\frac{1}{n!}\int_{-\infty}^{+\infty} dt_1\,V^{(n)}_{k_1\ldots k_n}(q^J_{cl}(t_1))\,\eta^{k_1}(t_1)\ldots\eta^{k_n}(t_1)\,,\end{aligned} \tag{3.148}$$

which when substituted into Eq.(3.135) and after changing the integration variables

$$q \to \eta\,, \qquad [\mathcal{D}q] = [\mathcal{D}\eta]\,, \tag{3.149}$$

enables us to find

$$\begin{aligned}\mathbf{Z[J]} &= \exp\left\{\frac{i}{\hbar}S^J[q^J_{cl}]\right\}\int[\mathcal{D}\eta] \\ &\times \exp\left\{\frac{i}{2\hbar}\int_{-\infty}^{+\infty} dt_1\int_{-\infty}^{+\infty} dt_2\,\eta^{k_1}(t_1)\,\Gamma_{k_1k_2}\left(q^J_{cl};t_1,t_2\right)\,\eta^{k_2}(t_2)\right. \\ &\left. - \frac{i}{\hbar}\sum_{n=3}^{\infty}\frac{1}{n!}\int_{-\infty}^{+\infty} dt_1\,V^{(n)}_{k_1\ldots k_n}(q^J_{cl}(t_1))\,\eta^{k_1}(t_1)\ldots\eta^{k_n}(t_1)\right\}.\end{aligned} \tag{3.150}$$

A quantum theory is defined by the set of *connected* Green functions whose generating functional, $\tilde{W}_0$, was introduced in Eq.(2.42). In accordance with Eqs.(2.42) and (3.134) we can write

$$e^{\frac{i}{\hbar}\tilde{W}_0[J|+\infty,-\infty]} = \frac{\mathbf{Z[J]}}{\mathbf{Z[J=0]}} . \tag{3.151}$$

We claim now that the stationary phase approximation gives rise to a series expansion of $\tilde{W}_0$ in powers of $\hbar$. To support this, we start by relabeling the integration variable

$$\eta \to \chi \equiv \frac{\eta}{\hbar^{\frac{1}{2}}} \tag{3.152}$$

which allows for casting Eq.(3.150) as

$$\begin{aligned}\mathbf{Z[J]} &= \exp\left\{\frac{i}{\hbar}S^J[q_{cl}^J]\right\}\int[\mathcal{D}\chi] \\ &\times \exp\left\{\frac{i}{2}\int_{-\infty}^{+\infty}dt_1\int_{-\infty}^{+\infty}dt_2\chi^{k_1}(t_1)\Gamma_{k_1k_2}\left(q_{cl}^J;t_1,t_2\right)\chi^{k_2}(t_2)\right. \\ &\left. -i\sum_{n=3}^{\infty}\frac{\hbar^{\frac{n}{2}-1}}{n!}\int_{-\infty}^{+\infty}dt_1V^{(n)}_{k_1\ldots k_n}(q_{cl}^J(t_1))\chi^{k_1}(t_1)\ldots\chi^{k_n}(t_1)\right\}. \end{aligned} \tag{3.153}$$

For an arbitrary function $V(q)$ the exact evaluation of the functional integral given above is beyond expectancy. What is left is to introduce a fictitious source u for the field χ in order to bring outside the functional integral all terms in the summation beginning at $n=3$, namely,

$$\begin{aligned}\mathbf{Z[J]} &= \exp\left\{\frac{i}{\hbar}S^J[q_{cl}^J]\right\} \\ &\times \exp\left\{-i\sum_{n=3}^{\infty}\frac{\hbar^{\frac{n}{2}-1}}{n!}\int_{-\infty}^{+\infty}dt_1V^{(n)}_{k_1\ldots k_n}(q_{cl}^J(t_1))\frac{1}{i}\frac{\delta}{\delta u_{k_1}(t_1)}\cdots\frac{1}{i}\frac{\delta}{\delta u_{k_n}(t_1)}\right\} \\ &\times\int[\mathcal{D}\chi]\exp\left\{\frac{i}{2}\int_{-\infty}^{+\infty}dt_1\int_{-\infty}^{+\infty}dt_2\chi^{k_1}(t_1)\Gamma_{k_1k_2}\left(q_{cl}^J;t_1,t_2\right)\chi^{k_2}(t_2)\right. \\ &\left.+i\int_{-\infty}^{+\infty}dt_1\chi^k(t_1)u_k(t_1)\right\}\Bigg|_{u=0} . \end{aligned} \tag{3.154}$$

The residual functional integral can be carried out at once. In accordance with Eq.(B.78) we end up with

$$\begin{aligned}
\mathbf{Z[J]} &= \exp\left\{\frac{i}{\hbar}S^J[q_{cl}^J]\right\}\left(\det\Gamma[q_{cl}^J]\right)^{-\frac{1}{2}} \\
&\times \exp\left\{-i\sum_{n=3}^{\infty}\frac{\hbar^{\frac{n}{2}-1}}{n!}\int_{-\infty}^{+\infty}dt_1 V^{(n)}_{k_1\ldots k_n}(q_{cl}^J(t_1))\frac{1}{i}\frac{\delta}{\delta u_{k_1}(t_1)}\cdots\frac{1}{i}\frac{\delta}{\delta u_{k_n}(t_1)}\right\} \\
&\times \exp\left\{-\frac{i}{2}\int_{-\infty}^{+\infty}d\tau'\int_{-\infty}^{+\infty}d\tau''\right. \\
&\times \left. u_k(\tau')\,\Delta^{kl}\left(q_{cl}^J;\tau',\tau''\right)\,u_l(\tau'')\right\}\Bigg|_{u=0}\,, \qquad (3.155)
\end{aligned}$$

where $\Delta\left(q_{cl}^J;\tau',\tau''\right)$ is the Green function of the operator $\Gamma\left(q_{cl}^J;\tau',\tau''\right)$, i.e.,

$$\int_{-\infty}^{+\infty} d\tau'\,\Gamma_{kl}\left(q_{cl}^J;\tau,\tau'\right)\,\Delta^{lm}\left(q_{cl}^J;\tau',\tau''\right) = \delta_k^{\,m}\,\delta\left(\tau-\tau''\right). \qquad (3.156)$$

Clearly, only terms involving an *even* number of functional derivatives $\frac{\delta}{\delta u_{k_n}(t_1)}$ contribute to $\mathbf{Z[J]}$. The same, of course, applies for $\mathbf{Z[J=0]}$. Thus, the perturbative solution under analysis is, in fact, a series expansion in powers of $\hbar$.

To facilitate the forthcoming calculations we shall make assumptions about the potential that do not imply in loosing generality. Only functions $V(q)$ which are analytic around $q=0$ and verify

$$V(0) = 0 \qquad (3.157)$$

will be considered. This implies that

$$S^{J=0}[q_{cl}^{J=0}] = 0\,, \qquad (3.158)$$

as it can be seen from Eqs.(3.130), (3.131), (3.143) and (3.157).

Next on the line of increasing powers of $\hbar$ is the quotient of determinants

$$\begin{aligned}
\left[\frac{\det\Gamma\left(q_{cl}^J\right)}{\det\Gamma\left(0\right)}\right]^{-\frac{1}{2}} &= \left\{\det\left[\Gamma^{-1}\left(0\right)\Gamma\left(q_{cl}^J\right)\right]\right\}^{-\frac{1}{2}} \\
= \left\{\det\left[\Delta\left(0\right)\Gamma\left(q_{cl}^J\right)\right]\right\}^{-\frac{1}{2}} &= \exp\left\{-\frac{1}{2}\ln\det\left[\Delta\left(0\right)\Gamma\left(q_{cl}^J\right)\right]\right\} \\
= \exp\left\{-\frac{1}{2}\mathrm{tr}\ln\left[\Delta\left(0\right)\Gamma\left(q_{cl}^J\right)\right]\right\}, & \qquad (3.159)
\end{aligned}$$

where $\Gamma(0)$ is the operator in Eq.(3.145) for $q^J_{cl}(t) = 0$, i.e.,

$$\Gamma_{k_1 k_2}(0;t,t') = \left[-M\,\delta_{k_1 k_2}\frac{d^2}{dt_1^2} - V^{(2)}_{k_1 k_2}(0)\right]\delta(t_1 - t_2)\,. \qquad (3.160)$$

Its Green function will be denoted by $\Delta^{k_1 k_2}(0;t_1,t_2)$. Furthermore, a relationship linking $\Gamma\left(q^J_{cl}\right)$ with $\Gamma\left(0\right)$ can be traced by starting from Eq.(3.145). It reads

$$\Gamma_{k_1 k_2}\left(q^J_{cl};t,t'\right) = \left[-M\delta_{k_1 k_2}\frac{d^2}{dt^2} - V^{(2)}_{k_1 k_2}(q^J_{cl}(t))\right]\delta\left(t - t'\right)$$
$$= \Gamma_{k_1 k_2}\left(0;t,t'\right) - \bar{V}^{(2)}_{k_1 k_2}(q^J_{cl}(t))\delta\left(t - t'\right)\,, \qquad (3.161)$$

where

$$\bar{V}^{(2)}_{k_1 k_2}(q^J_{cl}(t)) \equiv V^{(2)}_{k_1 k_2}(q^J_{cl}(t)) - \bar{V}^{(2)}_{k_1 k_2}(0)\,. \qquad (3.162)$$

Therefore,

$$\Delta\left(0\right)\Gamma\left(q^J_{cl}\right) = \Delta\left(0\right)\Gamma\left(0\right) - \Delta\left(0\right)\bar{V}^{(2)}(q^J_{cl})$$
$$= I - \Delta\left(0\right)\bar{V}^{(2)}(q^J_{cl}) \qquad (3.163)$$

and, moreover,

$$\ln\left[I - \Delta\left(0\right)\bar{V}^{(2)}(q^J_{cl})\right] = -\sum_{n=1}^{\infty}\frac{1}{n}\left[\Delta\left(0\right)\bar{V}^{(2)}(q^J_{cl})\right]^n\,. \qquad (3.164)$$

Summarizing,

$$\left[\frac{\det\Gamma\left(q^J_{cl}\right)}{\det\Gamma\left(0\right)}\right]^{-\frac{1}{2}} = -\frac{1}{2}\mathrm{tr}\ln\left[\Delta\left(0\right)\Gamma\left(q^J_{cl}\right)\right]$$
$$= -\frac{1}{2}\mathrm{tr}\ln\left[I - \Delta\left(0\right)\bar{V}^{(2)}(q^J_{cl})\right] = \sum_{n=1}^{\infty}\frac{1}{2n}\mathrm{tr}\left\{\left[\Delta\left(0\right)\bar{V}^{(2)}(q^J_{cl})\right]^n\right\}$$
$$= \sum_{n=1}^{\infty}\frac{1}{2n}\int_{-\infty}^{+\infty}dt_1\cdots\int_{-\infty}^{+\infty}dt_n\Delta^{k_1 j_1}\left(0;t_1,t_2\right)\bar{V}^{(2)}_{j_1 k_2}(q^J_{cl}(t_2))$$
$$\cdots\Delta^{k_n j_n}\left(0;t_n,t_1\right)\bar{V}^{(2)}_{j_n k_1}(q^J_{cl}(t_1))\,. \qquad (3.165)$$

We already have at hand all ingredients needed for computing the term of order $\hbar$ in the series expansion of $\tilde{W}_0[J|+\infty,-\infty]$. Indeed, from Eqs.(3.151), (3.155) and (3.165) it follows that

$$\begin{aligned}\tilde{W}_0[J|+\infty,-\infty] &= S^J[q^J_{cl}] + \frac{i}{2}\hbar \operatorname{tr}\ln\left[I - \Delta(0)\bar{V}^{(2)}(q^J_{cl})\right] + \mathcal{O}(\hbar^2)\\ &= S^J[q^J_{cl}] - i\hbar\sum_{n=1}^{\infty}\frac{1}{2n}\int_{-\infty}^{+\infty}dt_1\cdots\int_{-\infty}^{+\infty}dt_l\Delta^{k_1j_1}(0;t_1,t_2)\,\bar{V}^{(2)}_{j_1k_2}(q^J_{cl}(t_2))\\ &\cdots\Delta^{k_nj_n}(0;t_n,t_1)\,\bar{V}^{(2)}_{j_nk_1}(q^J_{cl}(t_1)) + \mathcal{O}(\hbar^2)\,.\end{aligned} \tag{3.166}$$

However, the existence of the Green function $\Delta(0)$ calls for additional restrictions on the potential. We shall demand $V^{(2)}(0)$ to be of the following form

$$V^{(2)}_{k_1k_2}(0) = \delta_{k_1k_2}\,C\,, \qquad C \neq 0\,. \tag{3.167}$$

A nonvanishing constant C secures that the operator $\Gamma(0)$ is nonsingular. In fact,

$$-M\,\delta_{k_1k_2}\,\frac{d^2}{dt^2}\delta(t-t')$$

does not possess a unique inverse when defined over the time interval $-\infty < t, t' < \infty$ (recall Eq.(3.67)). One can convince oneself that $C \neq 0$ secures the uniqueness of $\Delta(0)$ as well as the existence of the already mentioned energy gap. These are not independent features since they imply each other. The units of C are mass·second^{-2} and it can therefore be written as $C = M\omega^2$ which, as far as the operator $\Gamma(0)$ is concerned, bring us back to the case of the harmonic oscillator. Correspondingly

$$\Delta^{k_1k_2}(0;t,t') = \delta^{k_1k_2}\Delta_F(t,t')\,, \tag{3.168}$$

where $\Delta_F(t,t')$ is the Green function defined in Eq.(3.51). By substituting Eq.(3.168) into Eq.(3.166) we find

$$\begin{aligned}\tilde{W}_0[J|+\infty,-\infty] &= S^J[q^J_{cl}] + \frac{i}{2}\hbar \operatorname{tr}\ln\left[I - \Delta(0)\bar{V}^{(2)}(q^J_{cl})\right] + \mathcal{O}(\hbar^2)\\ &= S^J[q^J_{cl}] - i\hbar\sum_{n=1}^{\infty}\frac{1}{2n}\int_{-\infty}^{+\infty}dt_1\cdots\int_{-\infty}^{+\infty}dt_n\Delta_F(t_1,t_2)\,\bar{V}^{(2)}_{k_1k_2}(q^J_{cl}(t_2))\\ &\cdots\Delta_F(t_n,t_1)\,\bar{V}^{(2)}_{k_nk_1}(q^J_{cl}(t_1)) + \mathcal{O}(\hbar^2)\,.\end{aligned} \tag{3.169}$$

To get a deeper insight into the present systematics we shall specialize it for the case of a one dimensional system ($N = 1$) whose potential is the polynomial

$$V(q) = \frac{1}{2} M\omega^2 q^2 + \frac{g}{4!} q^4 . \tag{3.170}$$

Then, from Eqs.(3.170) and (3.162) we get

$$\bar{V}^{(2)}(q^J_{cl}(t)) = \frac{g}{2} \left(q^J_{cl}(t)\right)^2 , \tag{3.171}$$

which, along with Eq.(3.169), lead us to

$$\begin{aligned} \tilde{W}_0[J| + \infty, -\infty] &= S^J[q^J_{cl}] + \frac{i}{2} \hbar \operatorname{tr} \ln \left[I - \Delta(0) \bar{V}^{(2)}(q^J_{cl}) \right] + \mathcal{O}(\hbar^2) \\ &= S^J[q^J_{cl}] - i\hbar \sum_{n=1}^{\infty} \frac{1}{2n} \left(\frac{g}{2}\right)^n \int_{-\infty}^{+\infty} dt_1 \cdots \int_{-\infty}^{+\infty} dt_n \Delta_F(t_1, t_2) \left(q^J_{cl}(t_2)\right)^2 \\ &\cdots \Delta_F(t_n, t_1) \left(q^J_{cl}(t_1)\right)^2 + \mathcal{O}(\hbar^2) . \end{aligned} \tag{3.172}$$

Notice that all positive integer powers of g contribute to the correction of order $\hbar$ to $\tilde{W}_0[J|+\infty, -\infty]$. This correction can be graphically represented in terms of *Feynman diagrams* as shown in Fig.3.2. The rules that allow for writing down the analytic expression associated with a diagram are known as *Feynman rules.* Presently, they read:

1) a continuous internal line joining the instant of times t_k and t_j corresponds to a factor $\Delta_F(t_k - t_j)$,
2) a vertex at the instant of time t_k corresponds to a factor $\frac{g}{2} \left(q^J_{cl}(t_k)\right)^2$,
3) one is to integrate on t_k, $k = 1, 2, \ldots, n$, from $-\infty$ to $+\infty$.

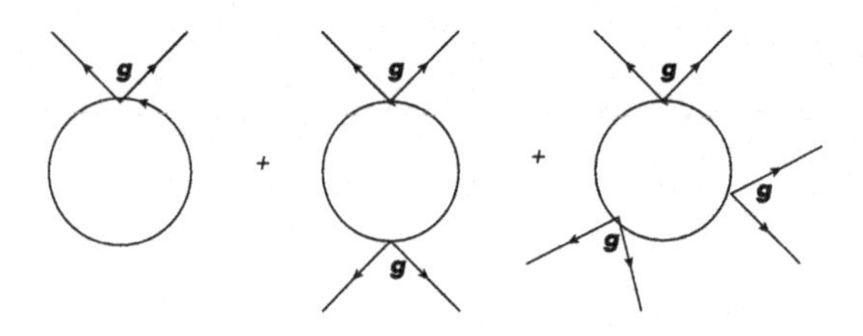

Fig. 3.2 Corrections of order $\hbar$.

For the model under analysis we can also work out the corrections of higher order in $\hbar$. The starting point is again Eq.(3.155) which presently undergoes several simplifications. In fact, all discrete indices collapse into a single one whereas the surviving vertices, besides $V^{(2)}$, are

$$V^{(3)}\left(q_{cl}^{J}(t)\right) = g q_{cl}^{J}(t)\,, \tag{3.173a}$$

$$V^{(4)}\left(q_{cl}^{J}(t)\right) = g\,. \tag{3.173b}$$

Thus, Eq.(3.155) reduces to

$$\begin{aligned}
\mathbf{Z[J]} &= \exp\left\{\frac{i}{\hbar}S^{J}[q_{cl}^{J}]\right\}\left(\det\Gamma[q_{cl}^{J}]\right)^{-\frac{1}{2}} \\
&\times \exp\left\{-\frac{i}{3!}\hbar^{\frac{1}{2}} g\int_{-\infty}^{+\infty} dt q_{cl}^{J}(t)\left[\frac{1}{i}\frac{\delta}{\delta u(t)}\right]^{3} - \frac{i}{4!}\hbar g\int_{-\infty}^{+\infty} dt\left[\frac{1}{i}\frac{\delta}{\delta u(t)}\right]^{4}\right\} \\
&\times \exp\left\{-\frac{i}{2}\int_{-\infty}^{+\infty} d\tau\int_{-\infty}^{+\infty} d\tau' u(\tau)\Delta\left(q_{cl}^{J};\tau,\tau'\right)u(\tau')\right\}\Bigg|_{u=0}.
\end{aligned} \tag{3.174}$$

We shall first look for the corrections of order $\hbar^2$. To that end (see Eq.(3.151)) we need to isolate the terms of order $\hbar$ in the right hand side of Eq.(3.174). We can check that

$$\begin{aligned}
&\exp\left\{-\frac{i}{3!}\hbar^{\frac{1}{2}} g\int_{-\infty}^{+\infty} dt q_{cl}^{J}(t)\left[\frac{1}{i}\frac{\delta}{\delta u(t)}\right]^{3} - \frac{i}{4!}\hbar g\int_{-\infty}^{+\infty} dt\left[\frac{1}{i}\frac{\delta}{\delta u(t)}\right]^{4}\right\} \\
&= 1 + \left(-\frac{i}{3!}\right)\hbar^{\frac{1}{2}} g\int_{-\infty}^{+\infty} dt q_{cl}^{J}(t)\left[\frac{1}{i}\frac{\delta}{\delta u(t)}\right]^{3} + \left(-\frac{i}{4!}\right)\hbar g\int_{-\infty}^{+\infty} dt\left[\frac{1}{i}\frac{\delta}{\delta u(t)}\right]^{4} \\
&+\frac{1}{2}\left(-\frac{i}{3!}\right)^{2}\hbar g^{2}\int_{-\infty}^{+\infty} dt q_{cl}^{J}(t)\left[\frac{1}{i}\frac{\delta}{\delta u(t)}\right]^{3}\int_{-\infty}^{+\infty} dt' q_{cl}^{J}(t')\left[\frac{1}{i}\frac{\delta}{\delta u(t')}\right]^{3} \\
&+\mathcal{O}\left(\hbar^{\frac{3}{2}}\right).
\end{aligned} \tag{3.175}$$

The number of derivatives associated with the term of order $\hbar^{\frac{1}{2}}$ is *odd* and such term can therefore be *disregarded*. The remaining functional derivatives are computed in the usual way. We arrive at

$$\begin{aligned}
&\left[\frac{1}{i}\frac{\delta}{\delta u(t)}\right]^{4}\exp\left\{-\frac{i}{2}\int_{-\infty}^{+\infty} d\tau\int_{-\infty}^{+\infty} d\tau' u(\tau)\Delta\left(q_{cl}^{J};\tau,\tau'\right)u(\tau')\right\}\Bigg|_{u=0} \\
&= 3(i)^{2}\Delta\left(q_{cl}^{J};t,t\right)\Delta\left(q_{cl}^{J};t,t\right)\,,
\end{aligned} \tag{3.176}$$

$$\begin{aligned}
&\left[\frac{1}{i}\frac{\delta}{\delta u(t')}\right]^{3}\left[\frac{1}{i}\frac{\delta}{\delta u(t)}\right]^{3}\exp\left\{-\frac{i}{2}\int_{-\infty}^{+\infty} d\tau\int_{-\infty}^{+\infty} d\tau' u(\tau)\Delta\left(q_{cl}^{J};\tau,\tau'\right)u(\tau')\right\}\Bigg|_{u=0} \\
&= 9(i)^{3}\Delta\left(q_{cl}^{J};t,t\right)\Delta\left(q_{cl}^{J};t,t'\right)\Delta\left(q_{cl}^{J};t',t'\right) \\
&+6(i)^{3}\Delta\left(q_{cl}^{J};t,t'\right)\Delta\left(q_{cl}^{J};t,t'\right)\Delta\left(q_{cl}^{J};t,t'\right).
\end{aligned} \tag{3.177}$$

By returning with these results into Eq.(3.175) and, afterwards, into Eq.(3.174) we find

$$\mathbf{Z[J]} = \exp\left\{\frac{i}{\hbar}S^J[q_{cl}^J]\right\}\left(\det\Gamma[q_{cl}^J]\right)^{-\frac{1}{2}}\exp\left\{\frac{i\hbar}{4!}\left[3g\int_{-\infty}^{+\infty}dt\Delta^2\left(q_{cl}^J;t,t\right)\right.\right.$$

$$+g^2\int_{-\infty}^{+\infty}dtq_{cl}^J(t)\int_{-\infty}^{+\infty}dt'q_{cl}^J(t')\left(3\Delta\left(q_{cl}^J;t,t\right)\Delta\left(q_{cl}^J;t,t'\right)\Delta\left(q_{cl}^J;t',t'\right)\right.$$

$$\left.\left.\left.+2\Delta\left(q_{cl}^J;t,t'\right)\Delta\left(q_{cl}^J;t,t'\right)\Delta\left(q_{cl}^J;t,t'\right)\right)\right]\right\}, \qquad (3.178)$$

which in the limit of vanishing fictitious sources reduces to

$$\mathbf{Z[J=0]} = \left(\det\Gamma[0]\right)^{-\frac{1}{2}}\exp\left\{\frac{i\hbar}{4!}3g\int_{-\infty}^{+\infty}dt\Delta_F^2\left(t,t\right)\right\}. \qquad (3.179)$$

After plugging back Eqs.(3.178) and (3.179) into Eq.(3.151) we end up with

$$\tilde{W}_0[J|+\infty,-\infty] = S^J[q_{cl}^J] + \frac{i}{2}\hbar\,\mathrm{tr}\ln\left[I-\Delta\left(0\right)\bar{V}^{(2)}(q_{cl}^J)\right]$$

$$+\frac{\hbar^2}{4!}\left\{3g\int_{-\infty}^{+\infty}dt\left[\Delta^2\left(q_{cl}^J;t,t\right)-\Delta_F^2\left(t,t\right)\right]\right.$$

$$+g^2\int_{-\infty}^{+\infty}dtq_{cl}^J(t)\int_{-\infty}^{+\infty}dt'q_{cl}^J(t')\left[3\Delta\left(q_{cl}^J;t,t\right)\Delta\left(q_{cl}^J;t,t'\right)\Delta\left(q_{cl}^J;t',t'\right)\right.$$

$$\left.\left.+2\Delta\left(q_{cl}^J;t,t'\right)\Delta\left(q_{cl}^J;t,t'\right)\Delta\left(q_{cl}^J;t,t'\right)\right]\right\}, \qquad (3.180)$$

where Eq.(3.172) has been taken into account. The corrections of order $\hbar^2$ are depicted in Fig.3.3. A dashed line corresponds to a factor $\Delta\left(q_{cl}^J;t,t'\right)$ while a continuous line corresponds, as previously stated, to $\Delta_F\left(t,t'\right)$. On the other hand, g and gq_{cl}^J are associated with four and three line vertices, respectively.

Higher order corrections to the generating functional $\tilde{W}_0[J|+\infty,-\infty]$ do not involve anything new except, of course, that their computation becomes algebraically cumbersome.

3.7 Effective action and effective potential

We introduce here the concepts of *effective action* and *effective potential.* They incorporate the quantum amendments to their classical counterparts.

We also compute the correction of order $\hbar$ to the potential in Eq.(3.170).

We denote by $q(t)$ the configuration

$$q(t) \equiv q[J|t] = \frac{\delta \tilde{W}_0[J]}{\delta J(t)}, \tag{3.181}$$

where, to simplify the writing, we have omitted in the argument of $\tilde{W}_0$ the initial $(-\infty)$ and final $(+\infty)$ instants of time. We shall continue to do so unless confusion arises. The equation

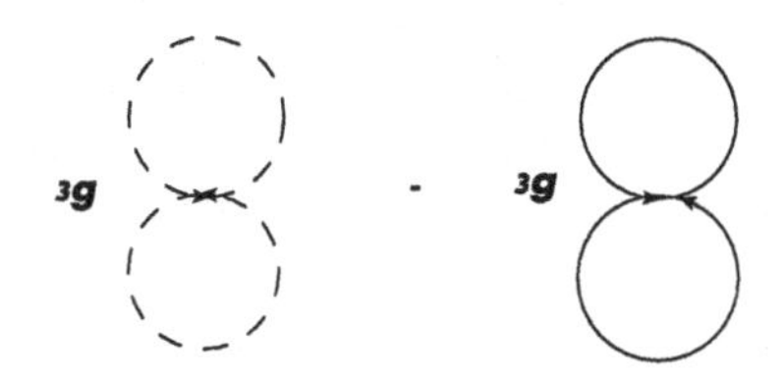

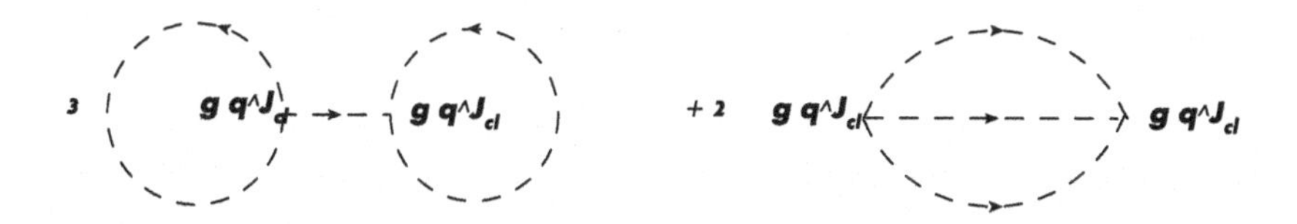

Fig. 3.3 Corrections of order $\hbar^2$.

$$q_c(t) = q[J_c|t] = \left. \frac{\delta \tilde{W}_0[J]}{\delta J(t)} \right|_{J=J_c}, \tag{3.182}$$

is, by assumption, invertible. This means that it can be solved to yield

$$J_c(t) = J[q_c|t]. \tag{3.183}$$

The *effective action* functional $\mathcal{A}[q]$ is defined by means of the Legendre transformation

$$\mathcal{A}[q] \equiv \left. \left[\tilde{W}_0[J] - \int_{-\infty}^{+\infty} dt J(t) q(t) \right] \right|_{J=J[q]}. \tag{3.184}$$

Now

$$\frac{\delta \mathcal{A}[q]}{\delta q(t)} = \frac{\delta}{\delta q(t)} \left\{ \left[\tilde{W}_0[J] - \int_{-\infty}^{+\infty} dt J(t) q(t) \right] \Big|_{J=J[q]} \right\}$$
$$= \frac{\delta \tilde{W}_0[J[q]]}{\delta q(t)} - \frac{\delta}{\delta q(t)} \int_{-\infty}^{+\infty} dt' J[q|t'] q(t')$$
$$= \int_{-\infty}^{+\infty} dt' \left\{ \left[\frac{\delta \tilde{W}_0[J]}{\delta J(t')} \Big|_{J=J[q]} - q(t') \right] \frac{\delta J[q|t']}{\delta q(t)} \right\} - J[q|t]. \quad (3.185)$$

In view of Eq.(3.181) the terms in brackets cancel and Eq.(3.185) reduces to

$$\frac{\delta \mathcal{A}[q]}{\delta q(t)} = -J[q|t]. \quad (3.186)$$

Hence, at the limit of vanishing fictitious sources $q(t)$ becomes a stationary configuration for the functional $\mathcal{A}[q]$. This explains why $\mathcal{A}[q]$ is being referred to as an *action*.

We look next for the relationship tying $q(t)$ with $q_{cl}^J(t)$. We invoke, for that purpose, Eqs.(3.181) and (3.172) which enable us to find

$$q(t) = \frac{\delta \tilde{W}_0[J]}{\delta J(t)} = \frac{\delta S^J[q_{cl}^J]}{\delta J(t)}$$
$$+ \frac{i}{2}\hbar \frac{\delta}{\delta J(t)} \left\{ \mathrm{tr} \ln \left[I - \Delta(0) \bar{V}^{(2)}(q_{cl}^J) \right] \right\} + \mathcal{O}(\hbar^2). \quad (3.187)$$

Moreover, in view of

$$S^J[q_{cl}^J] = S[q_{cl}^J] + \int_{-\infty}^{+\infty} dt q_{cl}^J(t) J(t) \quad (3.188)$$

we get

$$\frac{\delta S^J[q_{cl}^J]}{\delta J(t)} = \int_{-\infty}^{+\infty} dt' \left[\frac{\delta S[q]}{\delta q(t')} \Big|_{q=q_{cl}^J} + J(t') \right] \frac{\delta q_{cl}^J(t')}{\delta J(t)} + q_{cl}^J(t)$$
$$= q_{cl}^J(t), \quad (3.189)$$

where it was taken into account that the trajectory $q_{cl}^J(t)$ solves the equation of motion (3.139). Thus, Eq.(3.187) yields

$$q(t) - q_{cl}^J(t) = \mathcal{O}(\hbar). \quad (3.190)$$

We search for the consequences of this result at the level of the action $S^J[q]$. The Taylor expansion of $S^J[q]$ around the configuration $q = q^J_{cl}$ yields

$$
\begin{aligned}
S^J[q] &= S^J[q^J_{cl}] + \int_{-\infty}^{+\infty} dt \, \frac{\delta S^J[q]}{\delta q(t)}\bigg|_{q=q^J_{cl}} \left[q(t) - q^J_{cl}(t)\right] \\
&+ \frac{1}{2}\int_{-\infty}^{+\infty} dt \int_{-\infty}^{+\infty} dt' \, \frac{\delta^2 S^J[q]}{\delta q(t)\delta q(t')}\bigg|_{q=q^J_{cl}} \\
&\times \left[q(t) - q^J_{cl}(t)\right]\left[q(t) - q^J_{cl}(t')\right] + \cdots . \qquad (3.191)
\end{aligned}
$$

The second term in the right hand side of the latter equation drops out (see again Eq.(3.139)) while the third is, in accordance with Eq.(3.190), $\mathcal{O}(\hbar^2)$. Therefore,

$$
S^J[q] = S^J[q^J_{cl}] + \mathcal{O}(\hbar^2) . \qquad (3.192)
$$

Thus, Eq.(3.180) can be rearranged as follows

$$
\tilde{W}_0[J] = S^J[q] + \frac{i}{2}\,\hbar \,\mathrm{tr} \ln \left[I - \Delta(0)\, \bar{V}^{(2)}(q^J_{cl})\right] + \mathcal{O}(\hbar^2) , \qquad (3.193)
$$

which, along with Eq.(3.184), leads us to [9]

$$
\begin{aligned}
\mathcal{A}[q] &= S[q] + \frac{i}{2}\,\hbar \,\mathrm{tr} \ln \left[I - \Delta(0)\, \bar{V}^{(2)}(q)\right] + \mathcal{O}(\hbar^2) \\
&= S[q] - \frac{i}{2}\hbar \sum_{n=1}^{\infty} \frac{1}{n}\left(\frac{g}{2}\right)^n \int_{-\infty}^{+\infty} dt_1 \cdots \int_{-\infty}^{+\infty} dt_n \Delta_F(t_1, t_2)\,(q(t_2))^2 \\
&\cdots \Delta_F(t_n, t_1)\,(q(t_1))^2 + \mathcal{O}(\hbar^2) . \qquad (3.194)
\end{aligned}
$$

This is a relevant result within the present section. It gives the quantum correction of order $\hbar$ to the classical action.

We address now the problem of finding the *effective potential* ($\mathcal{V}(q)$) for the one dimensional model defined at the end of section 3.6. We begin by writing $\mathcal{V}(q)$ as the power series expansion

$$
\mathcal{V}(q) = V_0(q) + \hbar\, V_1(q) + \mathcal{O}(\hbar^2) , \qquad (3.195)
$$

[9]We remind the reader that $S^J[q] - \int_{-\infty}^{+\infty} dt q(t) J(t) = S[q]$.

where $V_0(q)$ is given at Eq.(3.170). Moreover, $\hbar V_1(q)$ is, up to an overall sign, what is left from the second term in the right hand side of Eq.(3.194) after setting [10]

$$q = \text{constant in time} \tag{3.196}$$

and factorizing without ambiguities a time divergent integral. Then, the computation of $V_1(q)$ starts from

$$-\frac{i}{2}\text{tr}\ln\left[I - \Delta(0)\bar{V}^{(2)}(q)\right]\Bigg|_{q=\text{constant}} = \frac{i}{2}\sum_{n=1}^{\infty}\frac{1}{n}\left(\frac{gq^2}{2}\right)^n$$
$$\times\int_{-\infty}^{+\infty}dt_1\cdots\int_{-\infty}^{+\infty}dt_n\Delta_F(t_1,t_2)\cdots\Delta_F(t_n,t_1)\,, \tag{3.197}$$

which in view of Eq.(3.63) can be rewritten as

$$-\frac{i}{2}\text{tr}\ln\left[I - \Delta(0)\bar{V}^{(2)}(q)\right]\Bigg|_{q=\text{constant}} = \int_{-\infty}^{+\infty}dt\Bigg\{\frac{i}{4\pi}\sum_{n=1}^{\infty}\frac{1}{n}\left(\frac{gq^2}{2M}\right)^n$$
$$\times\int_{-\infty}^{+\infty}dp\frac{1}{(p^2-\omega^2+i\epsilon)^n}\Bigg\}. \tag{3.198}$$

Hence, after disregarding the just mentioned divergent time integral we end up with

$$V^{(1)}(q) = \frac{i}{4\pi}\sum_{n=1}^{\infty}\frac{1}{n}\left(\frac{gq^2}{2M}\right)^n\int_{-\infty}^{+\infty}dp\frac{1}{(p^2-\omega^2+i\epsilon)^n}\,. \tag{3.199}$$

The integral on p will be performed by appealing to the *Wick rotation.* Consider the integral

$$\oint_{\mathbf{C}}dp\frac{1}{(p^2-\omega^2+i\epsilon)^n}\,, \tag{3.200}$$

where $\mathbf{C}$ is the closed path in the p complex plane depicted in Fig.3.4. The poles $p_{1,2}$ (see Eq.(3.62)) lie outside the integration contour. Then, from Cauchy theorem it follows that

[10] At the classical level, choosing $q =$ constant isolates the potential from the context of the action.

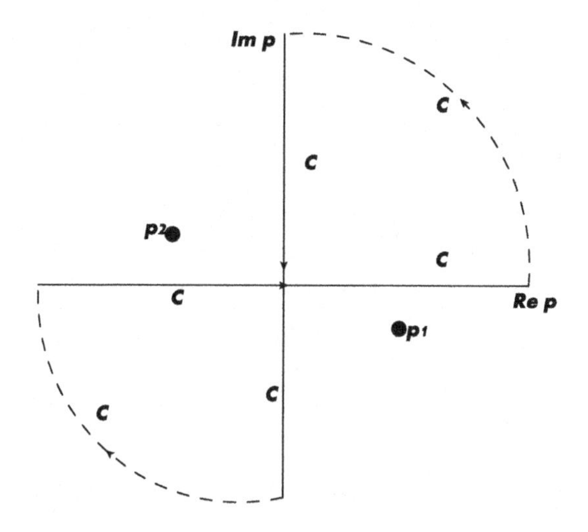

Fig. 3.4 Wick's rotation.

$$\oint_C dp \frac{1}{(p^2 - \omega^2 + i\epsilon)^n} = 0 \,. \tag{3.201}$$

Furthermore, when $|p| \to \infty$ the contribution to the integral arising from the dashed portion of the contour of integration vanishes. Hence, Eq.(3.201) reduces to

$$\int_{-\infty}^{+\infty} dp \frac{1}{(p^2 - \omega^2 + i\epsilon)^n} + \int_{+i\infty}^{-i\infty} dp \frac{1}{(p^2 - \omega^2 + i\epsilon)^n} = 0 \,, \tag{3.202}$$

which, in turn, implies that

$$\int_{-\infty}^{+\infty} dp \frac{1}{(p^2 - \omega^2 + i\epsilon)^n} = \int_{-i\infty}^{+i\infty} dp \frac{1}{(p^2 - \omega^2 + i\epsilon)^n} \,. \tag{3.203}$$

This is the previously referred Wick rotation. It establishes that certain improper integrals along the real and imaginary axes are equivalent. The change of the integration variable

$$p = ik \Longrightarrow dp = idk \,, \tag{3.204}$$

in the right hand side of Eq.(3.203) enables us to write

$$\int_{-\infty}^{+\infty} dp \frac{1}{(p^2 - \omega^2 + i\epsilon)^n} = (-1)^{-n}\, i \int_{-\infty}^{+\infty} dk \frac{1}{(k^2 + \omega^2)^n} \,. \tag{3.205}$$

Notice that the $i\epsilon$ device has been eliminated in the right hand side of the latter equation, since the corresponding integrand is free from obstructions along the integration path.

By replacing Eq.(3.205) into Eq.(3.199) we arrives at

$$\begin{aligned} V^{(1)}(q) &= -\frac{1}{4\pi}\int_{-\infty}^{+\infty} dk \sum_{n=1}^{\infty}\left[\frac{1}{n}\left(-\frac{gq^2}{2}\right)^n \frac{1}{(k^2+\omega^2)^n}\right] \\ &= \frac{1}{2\pi}\int_0^{+\infty} dk \ln\left(1+\frac{\frac{gq^2}{2M}}{k^2+\omega^2}\right) \\ &= \frac{\omega}{2}\left(\sqrt{1+\frac{gq^2}{2M\omega^2}}-1\right), \end{aligned} \tag{3.206}$$

where the integral on k was computed with the help of [Gradshteyn and Ryzhik (1980)]. Clearly, the large q behavior of the potential is modified by the quantum corrections [11].

[11]We bring to the attention of the reader the paper in [Coleman and Weinberg (1973)].

Chapter 4

Constrained systems

So far we have been dealing with regular systems. However, all interactions in nature are described by gauge theories which are systems with constraints. In this chapter, we focus on the quantization of such theories. In this sense, the functional formalism is the more appropriate tool for our purposes. After summarizing the Hamiltonian formulation of the classical dynamics of constrained systems, the construction of the generating functional of connected Green functions for gauge theories is then discussed.

4.1 Classical dynamics of constrained systems. Hamiltonian formulation

Let Γ be the phase space spanned by the variables q^i, p_i, $i = 1, \ldots, N$. The dynamics of a constrained system is generated by a Hamiltonian $h(q,p)$ along with the constraints

$$t_a(q,p) = 0\,, \qquad a = 1, \ldots, M\,, \qquad M < N\,. \tag{4.1}$$

For the time being we shall restrict ourselves to deal with first class systems, i.e., those possessing first class constraints (FCC's) only. In this case, both the Hamiltonian and the constraints obey the following involution algebra [Dirac (1964); Fradkin and Vilkovisky (1975, 1977); Faddeev (1970); Costa and Simões (1985)] [1]

$$[t_a(q,p)\,,\,t_b(q,p)]_{PB} = \sum_{c=1}^{M} C_{ab}^{c}(q,p)t_c(q,p)\,, \tag{4.2a}$$

$$[h(q,p)\,,\,t_a(q,p)]_{PB} = \sum_{b=1}^{M} B_a^b(q,p)t_b(q,p)\,. \tag{4.2b}$$

Here, $C_{ab}^{\;c}(q,p)$ and $B_a^{\;b}(q,p)$ are the *structure functions*. Notice that $C_{ab}^{\;c}(q,p)$ is antisymmetric under the exchange $a \leftrightarrow b$. When the $C_{ab}^{\;c}$'s are constants, the algebra in Eq.(4.2a) becomes a Lie algebra. This is the case, for instance, for the Maxwell and Yang-Mills fields. By assumption, the constraints are independent and irreducible [Faddeev (1970)] in the sense that they define a constraint surface Σ ($\Sigma \subset \Gamma$) of dimension $2N - M$. Also, $[f,h]_{PB}$ denotes the Poisson bracket in Γ.

The trajectories are the extremes of the action functional restricted to the constraint surface

$$S[q,p] = \int_{t_i}^{t_f} dt \left[\sum_{j=1}^{N} \dot{q}^j(t)\dot{p}_j(t) - h(q(t),p(t)) \right] \Bigg|_{q,p\in\Sigma} . \tag{4.3}$$

They are found by varying the equivalent action

$$S_E[q,p,\lambda] = \int_{t_i}^{t_f} dt \left[\sum_{j=1}^{N} p_j(t)\dot{q}^j(t) - h(q(t),p(t)) - \sum_{a=1}^{M} \lambda^a t_a \right] , \tag{4.4}$$

with respect to q, p and λ, where $\{\lambda^a | a = 1, \ldots, M\}$ is a set of Lagrange

[1]We bring to the reader's attention the following additional references on constrained systems: [Girotti (1990); Gitman and Tyutin (1990); Henneaux and Teitelboim (1992); Sundermeyer (1982); Sudarshan and Mukunda (1974)].

multipliers. We find that

$$\left.\frac{\delta S_E[q,p,\lambda]}{\delta p_j}\right|_{q,p\in\Sigma} = 0 \Longrightarrow \left.\dot{q}^j\right|_{q,p\in\Sigma}$$

$$= \left.\left[q^j, h(q,p)\right]_{PB}\right|_{q,p\in\Sigma} + \sum_{a=1}^{M} \lambda^a \left.\left[q^j, t_a(q,p)\right]_{PB}\right|_{q,p\in\Sigma}$$

$$= \left.\left[q^j, h_E(q,p)\right]_{PB}\right|_{q,p\in\Sigma}, \quad j = 1,\ldots,N, \tag{4.5a}$$

$$\left.\frac{\delta S_E[q,p,\lambda]}{\delta q^j j}\right|_{q,p\in\Sigma} = 0 \Longrightarrow \left.\dot{p}_j\right|_{q,p\in\Sigma}$$

$$= \left.[p_j, h(q,p)]_{PB}\right|_{q,p\in\Sigma} + \sum_{a=1}^{M} \lambda^a \left.[p_j, t_a(q,p)]_{PB}\right|_{q,p\in\Sigma}$$

$$= \left.[p_j, h_E(q,p)]_{PB}\right|_{q,p\in\Sigma}, \quad j = 1,\ldots,N, \tag{4.5b}$$

$$\left.\frac{\delta S_E[q,p,\lambda]}{\delta \lambda^a}\right|_{q,p\in\Sigma} = 0 \Longrightarrow$$

$$\left.t_a(q,p)\right|_{q,p\in\Sigma} = 0, \quad a = 1,\ldots,M, \tag{4.5c}$$

where

$$h_E(q,p) \equiv h(q,p) + \sum_{a=1}^{M} \lambda^a t_a(q,p) \tag{4.6}$$

is the *extended Hamiltonian* [Dirac (1964)]. All Poisson brackets are to be evaluated in Γ and only afterwards we set $q,p \in \Sigma$. Subject to this rule $[f(q,p), t_a(q,p)]_{PB}$ turns out to be a well-defined operation. Accordingly, the constraint equations (4.5c) cannot be used before working out all Poisson brackets [Dirac (1964)]. One can remember about these rules by introducing the sign of weak equality ($\approx$) and rewriting the equations of motion and constraint equations as

$$\begin{aligned}
\dot{q}^j &\approx \left[q^j, h(q,p)\right] + \sum_{a=1}^{M} \lambda^a \left[q^j, t_a\right]_{PB} \\
&\approx \left[q^j, h_E(q,p)\right], \quad j = 1, \ldots, N, && (4.7a) \\
\dot{p}_j &\approx \left[p_j, h(q,p)\right] + \sum_{a=1}^{M} \lambda^a \left[p_j, t_a\right]_{PB} \\
&\approx \left[p_j, h_E(q,p)\right], \quad j = 1, \ldots, N, && (4.7b) \\
t_a(q,p) &\approx 0, \quad a = 1, \ldots, M. && (4.7c)
\end{aligned}$$

Hence, the development in time of an arbitrary function $g(q(t), p(t), t)$ is given by

$$\begin{aligned}
\dot{g}(q,p,t) &\approx \left[g(q,p,t), h_E(q,p)\right]_{PB} + \frac{\partial g(q,p,t)}{\partial t} \\
&= \left[g(q,p,t), h(q,p)\right]_{PB} + \sum_{a=1}^{M} \lambda^a \left[g(q,p,t), t_a\right]_{PB} \\
&+ \frac{\partial g(q,p,t)}{\partial t}. && (4.8)
\end{aligned}$$

Now, consistency requires the persistency in time of the constraints. To see whether this holds true we use Eq.(4.8) for $g = t_a$, i.e.,

$$\begin{aligned}
\dot{t}_a(q,p) &\approx \left[t_a(q,p), h_E(q,p)\right]_{PB} \approx \left[t_a(q,p), h(q,p)\right]_{PB} \\
&+ \sum_{b=1}^{M} \lambda^b \left[t_a(q,p), t_b(q,p)\right]_{PB} \\
&\approx \sum_{c=1}^{M} \left[-B_a^{\ c}(q,p) + \lambda^b C_{ab}^{\ \ c}(q,p)\right] t_c(q,p) \approx 0, && (4.9)
\end{aligned}$$

as required. Observe that Eqs.(4.2) are at the root of this result. Moreover,

$$\dot{t}_a(q,p)\Big|_{q,p\in\Sigma} = 0 \tag{4.10}$$

holds *for any value of the* λ's. In other words, the λ's are not determined from the dynamics and, therefore, an initial data gives rise to a set of classical trajectories parametrized by the functions $\lambda^a(t), a = 1, \ldots, M$.

This reflects the fact that the action in Eq.(4.3) remains invariant under the infinitesimal transformations

$$\delta q^j = \left[q^j, G\right]_{PB}, \tag{4.11a}$$

$$\delta p_j = \left[p_j, G\right]_{PB}, \tag{4.11b}$$

which are generated by

$$G(q,p,t) \equiv \sum_{a=1}^{M} \epsilon^a(t) t_a(q,p), \tag{4.12}$$

where the $\epsilon^a(t)$'s are arbitrary infinitesimal parameters. The same applies for the action in Eq.(4.4) provided the transformations in Eq.(4.11) are supplemented by [Fradkin and Vilkovisky (1975, 1977); Faddeev (1970)]

$$\delta\lambda^a = \dot{\epsilon}^a - \sum_{b=1}^{M} B^a_b \epsilon^b + \sum_{b,c=1}^{M} C^a_{bc} \epsilon^b \lambda^c . \tag{4.13}$$

Thus, at a fixed time there are points belonging to different trajectories that represent the *same physical state of the system.* This amounts to say that the constraint surface Σ divides out into *equivalence classes*, all points within a given class are physically equivalent and interconnected by transformations generated by G with time independent parameters [2]. These are the so-called *gauge transformations.*

Nevertheless, we emphasize that the time evolution of quantities verifying

$$[f(q,p)\,,\, t_a(q,p)] = \sum_{a=1}^{M} D_a^{\ b}(q,p) t_b(q,p) \approx 0\,, \quad \forall\, a = 1, \ldots, M\,, \tag{4.14}$$

turns out to be unique because the Lagrange multipliers do not enter the corresponding equations of motion. They are known as first class or *gauge invariant* quantities. In particular, Eq.(4.2b) secures that the Hamiltonian is first class.

We address next to the problem of parametrizing the space of classes. It amounts to select one and only one configuration out from each set of *gauge equivalent histories.* This is done by calling upon a new set of functions

[2] For regular systems the trajectory is unique and each of its points represents a physical state.

$$u^a(q,p) = 0\,, \quad a = 1, \ldots, M\,, \tag{4.15}$$

known as *subsidiary or gauge conditions.* They must be admissible in the sense of selecting *at least* one history from each equivalence class. Furthermore, they must eliminate the gauge freedom completely and therefore be fulfilled *at the most* by one of the histories in each equivalence class. It has been shown [Fradkin and Vilkovisky (1975, 1977); Faddeev (1970)] that the second requirement demands the choosing of the gauge conditions such that

$$\det \| \, [u^a \,, t_b]_{PB} \, \| \neq 0\,. \tag{4.16}$$

This amounts to saying that the gauge conditions along with the FCC's give rise to a set of *second class constraints* [Faddeev (1970)] [3].

By adding the gauge conditions the *number of constraints doubles.* Then, the motion takes place in the reduced or *physical phase space* Γ^* with dimensions $2N - M - M = 2(N - M)$. As for the dynamics, it becomes specified by the *complete* action

$$\begin{aligned} S_C[q,p,\lambda,\chi] = \int_{t_i}^{t_f} dt \Bigg[\sum_{j=1}^{N} p_j(t)\dot{q}^j(t) - h(q(t),p(t)) \\ - \sum_{a=1}^{M} \lambda^a t_a - \sum_{a=1}^{M} \chi_a u^a \Bigg]\,, \end{aligned} \tag{4.17}$$

where $\chi_a, a = 1, \ldots, M$, is a new set of Lagrange multipliers. This last expression can be cast

$$\begin{aligned} S_C[q,p,\lambda,\chi] = \int_{t_i}^{t_f} dt \Bigg[\sum_{j=1}^{N} p_j(t)\dot{q}^j(t) - h(q(t),p(t)) \\ - \sum_{\alpha=1}^{2M} \xi^\alpha \psi_\alpha \Bigg]\,, \end{aligned} \tag{4.18}$$

with

$$\xi^\alpha \equiv \left[\lambda^a, \chi_a \right] \tag{4.19}$$

[3] What we have in Eq.(4.16) is the determinant of an $M \times M$ matrix known as the Faddeev-Popov matrix [Faddeev (1970); Faddeev and Popov (1967)].

and

$$\psi_\alpha \equiv \begin{bmatrix} t_a \\ u^a \end{bmatrix} . \tag{4.20}$$

Let us now look for the Hamilton equations of motions deriving from $S_C[q,p,\lambda,\chi]$. Through the usual steps we get

$$\dot{q}^j \approx \left[q^j, h(q,p)\right]_{PB} + \sum_{\alpha=1}^{2M} \xi^\alpha \left[q^j, \psi_\alpha(q,p)\right]_{PB} , \tag{4.21a}$$

$$\dot{p}_j \approx \left[p_j, h(q,p)\right]_{PB} + \sum_{\alpha=1}^{2M} \xi^\alpha \left[p_j, \psi_\alpha(q,p)\right]_{PB} , \tag{4.21b}$$

whereas a regular function $g(q,p,t)$ develops in time in accordance with

$$\dot{g}(q,p,t) \approx \left[g, h(q,p)\right]_{PB} + \sum_{\alpha=1}^{2M} \xi^\alpha \left[g, \psi_\alpha(q,p)\right]_{PB} + \frac{\partial g}{\partial t} . \tag{4.22}$$

However, unlike it happens before, the persistence in time of the constraints does not leave us with undetermined Lagrange multipliers. Indeed, by starting from

$$\begin{aligned} \dot{\psi}_\alpha(q,p) &\approx \left[\psi_\alpha(q,p), h(q,p)\right]_{PB} + \sum_{\beta=1}^{2M} \xi^\beta \left[\psi_\alpha(q,p), \psi_\beta(q,p)\right]_{PB} \\ &\approx 0 \end{aligned} \tag{4.23}$$

we find

$$\sum_{\beta=1}^{2M} Q_{\alpha\beta}\, \xi^\beta = - \left[\psi_\alpha(q,p), h(q,p)\right]_{PB} \Big|_{q,p\in\Gamma^\star} , \tag{4.24}$$

where

$$Q_{\alpha\beta} \equiv \left[\psi_\alpha(q,p), \psi_\beta(q,p)\right]_{PB} \Big|_{q,p\in\Gamma^\star} \tag{4.25}$$

are the elements of an antisymmetric $2M \times 2M$ matrix Q. Since Eq.(4.16) implies that

$$\det Q \neq 0 \tag{4.26}$$

Q turns out to be a non-singular matrix and, as a consequence, the solving of Eq.(4.24) yields the following *unique solution* ($R \equiv Q^{-1}$)

$$\xi^\alpha = -\sum_{\beta=1}^{2M} R^{\alpha\beta} \left[\psi_\beta(q,p), h(q,p)\right]_{PB} . \tag{4.27}$$

Moreover, Eqs.(4.21) boil down to

$$\dot{q}^j \approx \left[q^j, h(q,p)\right]_{DB} , \tag{4.28a}$$
$$\dot{p}_j \approx \left[p_j, h(q,p)\right]_{DB} , \tag{4.28b}$$

or, equivalently,

$$\dot{q}^j\bigg|_{q,p\in\Gamma^\star} = \left[q^j, h(q,p)\right]_{DB}\bigg|_{q,p\in\Gamma^\star} , \tag{4.29a}$$
$$\dot{p}_j\bigg|_{q,p\in\Gamma^\star} = \left[p_j, h(q,p)\right]_{DB}\bigg|_{q,p\in\Gamma^\star} . \tag{4.29b}$$

At last, the variation of the action in Eq.(4.18) with respect to ξ^α gives rise to

$$\psi_\alpha(q,p) \approx 0 \Longrightarrow \psi_\alpha(q,p)\bigg|_{q,p\in\Gamma^\star} = 0 . \tag{4.30}$$

A new bracket, the Dirac bracket (DB) [Dirac (1964)], was introduced in Eqs.(4.28). For two arbitrary functions $f(q,p)$ and $h(q,p)$ it is defined as

$$\left[f(q,p), h(q,p)\right]_{DB} \equiv \left[f(q,p), h(q,p)\right]_{PB} - \sum_{\alpha\beta=1}^{2M} \left[f, \psi_\alpha\right]_{PB} R^{\alpha\beta} \left[\psi_\beta, g\right]_{PB} . \tag{4.31}$$

We can verify that within the DB algebra the constraints and gauge conditions hold as strong identities. Therefore,

$$\dot{\psi}_\alpha(q,p) \approx \left[\psi_\alpha(q,p), h(q,p)\right]_{DB} = 0 , \tag{4.32}$$

as required by consistency.

In [Fradkin and Vilkovisky (1975, 1977)] it is shown that it is possible to get rid of the constraints and the gauge conditions by introducing canonical coordinates in $\Gamma^\star$,

$$q^{\star L}\,,\, p^\star_L\,, \qquad L = 1, \ldots, N - M\,, \tag{4.33}$$

and then finding the $2N$ functions

$$q^j = q^j(q^\star\,,\, p^\star)\,, \quad j = 1, \ldots, N\,, \tag{4.34a}$$

$$p_j = p_j(q^\star\,,\, p^\star)\,, \quad j = 1, \ldots, N\,, \tag{4.34b}$$

such that

$$\psi_\alpha(q,p)\Big|_{q,p\in\Gamma^\star} = \psi_\alpha\left(q(q^\star,p^\star)\,,\, p(q^\star,p^\star)\right) = 0, \;\; \alpha = 1, \ldots, 2M \tag{4.35}$$

and

$$\int_{t_i}^{t_f} dt \sum_{j=1}^{N} p_j \dot{q}^j\Big|_{q,p\in\Gamma^\star} = \int_{t_i}^{t_f} dt \sum_{L=1}^{N-M} p^\star_L \dot{q}^{\star L}\,. \tag{4.36}$$

Moreover, the Hamilton equations of motions for the variables spanning $\Gamma^\star$ have been shown to read [Fradkin and Vilkovisky (1975, 1977)]

$$\dot{q}^{\star L} = \left[q^{\star L}, h_{phys}(q^\star,p^\star)\right]_{PB} = \frac{\partial h_{phys}(q^\star,p^\star)}{\partial p^\star_L}\,, \tag{4.37a}$$

$$\dot{p}^\star_L = [p^\star_L, h_{phys}(q^\star,p^\star)]_{PB} = -\frac{\partial h_{phys}(q^\star,p^\star)}{\partial q^{\star L}}\,, \tag{4.37b}$$

where

$$h_{phys}(q^\star,p^\star) \equiv h(q,p)\Big|_{q,p\in\Gamma^\star} = h\left(q(q^\star,p^\star)\,,\, p(q^\star,p^\star)\right) \tag{4.38}$$

is the *physical* Hamiltonian. From [Fradkin and Vilkovisky (1975, 1977)] we also get that

$$[f(q,p), h(q,p)]_{DB}\Big|_{q,p\in\Gamma^\star} = [f^\star, h_{phys}]_{PB} \tag{4.39}$$

where

$$f^\star \equiv f(q,p)\Big|_{q,p\in\Gamma^\star} = f\left(q(q^\star,p^\star)\,,\,p(q^\star,p^\star)\right)\,. \tag{4.40}$$

It is also worth mentioning that the $2N$ initial conditions $q(t_i), p(t_i)$, needed for solving Eqs.(4.29), are to be connected with the $2(N-M)$ initial conditions $q^\star(t_i), p^\star(t_i)$, needed for solving Eqs.(4.37), through the relationships (4.34) specialized for the initial time, i.e.,

$$q^j(t_i)\Big|_{q,p\in\Gamma^\star} = q^j\left(q^\star(t_i), p^\star(t_i)\right)\,, \quad j = 1,\ldots,N\,, \tag{4.41a}$$

$$p_j(t_i)\Big|_{q,p\in\Gamma^\star} = p_j\left(q^\star(t_i), p^\star(t_i)\right)\,, \quad j = 1,\ldots,N\,. \tag{4.41b}$$

As for the uniqueness of the dynamics, we state without proof that a change of the gauge conditions corresponds to a canonical transformation in $\Gamma^\star$ [Faddeev (1970)].

4.2 Functional quantization of constrained systems

What we have in $\Gamma^\star$ is a regular system for which the results in section 3.5 apply. Let us focus on the denominator of Eq.(3.136) which in accordance with Eq.(3.137) reads

$$\begin{aligned}\mathbf{Z}[\mathbf{J}{=}0\,,\mathbf{K}{=}0] &= \int [\mathcal{D}q^\star]\int [\mathcal{D}p^\star] \\ &\times \exp\left\{\frac{i}{\hbar}\int_{-\infty}^{+\infty} dt\,\left[p^\star_L(t)\dot{q}^{\star L}(t) - h_{phys}(q^\star(t),p^\star(t))\right]\right\}\,,\end{aligned} \tag{4.42}$$

where

$$[\mathcal{D}q^\star] \equiv \prod_{L=1}^{N-M}\left[\mathcal{D}q^{\star L}\right]\,, \tag{4.43a}$$

$$[\mathcal{D}p^\star] \equiv \prod_{L=1}^{N-M}\left[\mathcal{D}p^\star_L\right]\,. \tag{4.43b}$$

The task consists of writing $\mathbf{Z}[\mathbf{J}{=}0\,,\mathbf{K}{=}0]$ in terms of the constrained variables q and p. Such expression was found in [Faddeev (1970); Faddeev and Popov (1967)] and reads

$$\mathbf{Z[J{=}0\,,K{=}0]} = \int [\mathcal{D}q] \int [\mathcal{D}p] \left(\prod_{\alpha=1}^{2M} \delta\left[\psi_\alpha\right] \right) [\det Q]^{\frac{1}{2}}$$
$$\times \exp\left\{ \frac{i}{\hbar} \int_{-\infty}^{+\infty} dt \left[p_j(t)\dot{q}^j(t) - h(q(t),p(t)) \right] \right\} . \tag{4.44}$$

Here, $\delta\left[\psi_\alpha\right]$ is the generalized Dirac delta *functional* defined in Eq.(B.89), while $[\det Q]$ denotes the *functional* determinant of the matrix defined in Eq.(4.25). The structure of the functional integral in the right hand side of Eq.(4.44) differs from the correspondent for regular systems by modifications in the functional measure. Indeed, instead of $[\mathcal{D}q]\,[\mathcal{D}p]$ we now have

$$[\mathcal{D}\mu] \equiv [\mathcal{D}q]\,[\mathcal{D}p] \left(\prod_{\alpha=1}^{2M} \delta\left[\psi_\alpha\right] \right) [\det Q]^{\frac{1}{2}} \,. \tag{4.45}$$

The delta functions bring into the path integral the restrictions imposed by the constraints and the gauge conditions. The reader is to recall that, according to Eq.(4.20),

$$\prod_{\alpha=1}^{2M} \delta\left[\psi_\alpha\right] = \prod_{a=1}^{M} \delta[t_a]\,\delta[u^a] \,. \tag{4.46}$$

What remains to be understood is the need for the Faddeev-Popov functional determinant $\left([\det Q]^{\frac{1}{2}}\right)$.

We show next that Eqs.(4.42) and (4.44) are, in fact, equivalent. We follow, for this purpose, [Faddeev (1970)] and choose the gauge conditions such that

$$[u^a(q,p), u^b(q,p)]_{PB} = 0 \,, \qquad a\,,b = 1,\ldots,M \,. \tag{4.47}$$

This, along with Eq.(4.2a), leads us to

$$\det Q = \left(\det \|\, [u^a\,,\,t_b]_{PB}\,\|\right)^2 . \tag{4.48}$$

Hence, Eq.(4.44) can be written as

$$\mathbf{Z[J{=}0\,,K{=}0]}$$
$$= \int [\mathcal{D}q] \int [\mathcal{D}p] \left(\prod_{a=1}^{M} \delta[u^a]\,\delta[t_a] \right) [\det \|\, [u^a\,,\,t_b]_{PB}\,\|]$$
$$\times \exp\left\{ \frac{i}{\hbar} \int_{-\infty}^{+\infty} dt \left[p_j(t)\dot{q}^j(t) - h(q(t),p(t)) \right] \right\} . \tag{4.49}$$

By means of a canonical transformation we go onto a new set of phase space variables that is chosen such that

$$u^a(q,p) = p_a\,, \qquad a = 1,\ldots,M\,, \tag{4.50}$$

where $p_a, a = 1,\ldots,M$, is a subset of the new momenta. We denote $q^a, a = 1,\ldots,M$, the corresponding subset of canonical conjugate coordinates. The remaining part $(\Gamma^\star)$ of the phase space is spanned by the canonical variables $q^\star$, $p^\star$. In terms of the new variables, Eq.(4.16) takes the following form

$$\det\left\|\frac{\partial t_a}{\partial q^b}\right\| \neq 0 \tag{4.51}$$

while Eq.(4.49) goes into

$$\begin{aligned}
&\mathbf{Z[J{=}0\,,K{=}0]}\\
&= \int[\mathcal{D}q]\int[\mathcal{D}p]\left(\prod_{a=1}^{M}\delta[p_a]\,\delta[t_a]\right)\left[\det\left\|\frac{\partial t_a}{\partial q^b}\right\|\right]\\
&\times\,\exp\left\{\frac{i}{\hbar}\int_{-\infty}^{+\infty}dt\,\left[p_j(t)\dot{q}^j(t)\;-\;h(q(t),p(t))\right]\right\}.
\end{aligned} \tag{4.52}$$

Notice that Eq.(4.51) secures the solvability of Eqs.(4.1). We can, therefore, take

$$p_a = p_a(q^\star,p^\star) = 0\,, \qquad a = 1,\ldots,M\,, \tag{4.53a}$$

$$q^a = q^a(q^\star,p^\star)\,, \qquad a = 1,\ldots,M\,, \tag{4.53b}$$

as the defining equations for $\Gamma^\star \subset \Gamma$.

Now,

$$[\mathcal{D}q]\,[\mathcal{D}p] = \left(\prod_{a=1}^{N-M}[\mathcal{D}q^a]\right)\left(\prod_{a=1}^{N-M}[\mathcal{D}p_a]\right)[\mathcal{D}q^\star]\,[\mathcal{D}p^\star]\,. \tag{4.54}$$

Moreover

$$\begin{aligned}
t_a(q,p) &= t_a(q,p)\Big|_{q,p\in\Gamma^\star} + \left(q^b - q^b(q^\star,p^\star)\right)\frac{\partial t_a}{\partial q^b}\Big|_{q,p\in\Gamma^\star}\\
&+ \left(p_b - p_b(q^\star,p^\star)\right)\frac{\partial t_a}{\partial p_b}\Big|_{q,p\in\Gamma^\star} + \mathcal{O}\left(\Delta q\right)^2 + \mathcal{O}\left(\Delta p\right)^2\,,
\end{aligned} \tag{4.55}$$

which in view of Eq.(4.53a) and of

$$t_a(q,p)\Big|_{q,p\in\Gamma^\star} = 0 \tag{4.56}$$

reduces to

$$\begin{aligned} t_a(q,p) &= \left(q^b - q^b(q^\star,p^\star)\right) \frac{\partial t_a}{\partial q^b}\Big|_{q,p\in\Gamma^\star} + p_b \frac{\partial t_a}{\partial p_b}\Big|_{q,p\in\Gamma^\star} \\ &+ \mathcal{O}\left(\Delta q\right)^2 + \mathcal{O}\left(\Delta p\right)^2 . \end{aligned} \tag{4.57}$$

Thus,

$$\delta(p_a)\,\delta(t_a) \;=\; \delta(p_a)\,\delta\left\{ \left(q^b - q^b(q^\star,p^\star)\right) \frac{\partial t_a}{\partial q^b}\Big|_{q,p\in\Gamma^\star} \right\} \tag{4.58}$$

and, therefore,

$$\delta[p_a]\,\delta[t_a] \;=\; \delta[p_a]\,\delta[(q^a - q^a(q^\star,p^\star)]\left[\det\left|\left|\frac{\partial t_a}{\partial q^b}\right|\right|\right]^{-1} \tag{4.59}$$

where Eq.(B.97) was taken into account.

By returning with Eqs.(4.54) and (4.59) into Eq.(4.52) and after performing the integrals with respect to q^a and p_a we arrive at Eq.(4.42), which substantiates the validity of Eq.(4.44).

However, this is not the end of the story because the physically meaningful quantities must be gauge independent. This should be the case for **Z[J=0 , K=0]** since it is proportional to the propagator. We are, then, required to show that **Z[J=0 , K=0]** is unaffected by the change in the gauge conditions

$$u^a(q,p) = u^a(q,p) + \bar{\delta} u^a(q,p)\,, \qquad a = 1,\ldots,M\,. \tag{4.60}$$

This will be done by proving that (4.60) equals the outcome from the infinitesimal canonical transformation below

$$\begin{aligned} & q^i(t) \to \tilde{q}^i(t) = q^i(t) \,+\, \bar{\delta} q = q^i(t) \,+\, [q^i(t), \Phi]_{PB} \\ & = q^i(t) \,+\, \frac{\partial \Phi}{\partial p_i(t)}\,, \end{aligned} \tag{4.61a}$$

$$\begin{aligned} & p_i(t) \to \tilde{p}_i(t) = p_i(t) \,+\, \bar{\delta} p_i = p_i(t) \,+\, [p_i(t), \Phi]_{PB} \\ & = p_i(t) \,-\, \frac{\partial \Phi}{\partial q^i(t)}\,, \end{aligned} \tag{4.61b}$$

with the following generating function

$$\Phi \equiv \sum_{a=1}^{M} \epsilon^a \, t_a(q,p) \,. \tag{4.62}$$

The point of departure is $\mathbf{Z[J{=}0\,,K{=}0]}$ written in terms of the transformed variables, i.e.,

$$\begin{aligned}
&\mathbf{Z[J{=}0\,,K{=}0]} \\
&= \int [\mathcal{D}\tilde{q}] \int [\mathcal{D}\tilde{p}] \left(\prod_{a=1}^{M} \delta[u^a(\tilde{q},\tilde{p})] \, \delta[t_a(\tilde{q},\tilde{p})] \right) [\det \| \, [u^a(\tilde{q},\tilde{p}) \, , \, t_b(\tilde{q},\tilde{p})]_{PB} \, \|] \\
&\times \exp \left\{ \frac{i}{\hbar} \int_{-\infty}^{+\infty} dt \, \left[\tilde{p}_j(t) \dot{\tilde{q}}^j(t) \; - \; h(\tilde{q},\tilde{p}) \right] \right\} .
\end{aligned} \tag{4.63}$$

To find the change in the measure,

$$[\mathcal{D}\tilde{q}] \, [\mathcal{D}\tilde{p}] \; = \; \Omega \; [\mathcal{D}q] \, [\mathcal{D}p] \; , \tag{4.64}$$

we must compute the Jacobian of the transformation (Ω). It is the modulus of the functional determinant of the matrix $\|\Omega\|$ whose elements $(\langle r,t|\Omega|s,t'\rangle)$, labeled by discrete ($r,s = 1,\ldots,2N$) and continuous (t,t') indices, can be found through Eqs.(4.61). They are

$$\langle r,t|\Omega|s,t'\rangle = \frac{\delta \tilde{q}^i(t)}{\delta q^j(t')} = \delta_i^{\;j} \, \delta(t-t') + \frac{\partial^2 \Phi}{\partial p_i(t) \partial q^j(t)} \, \delta(t-t') \, , \qquad r=i, \; s=j \, , \tag{4.65a}$$

$$\langle r,t|\Omega|s,t'\rangle = \frac{\delta \tilde{q}^i(t)}{\delta p_j(t')} = \frac{\partial^2 \Phi}{\partial p_i(t) \partial p_j(t)} \, \delta(t-t') \, , \qquad r=i, \; s=j+N \, , \tag{4.65b}$$

$$\langle r,t|\Omega|s,t'\rangle = \frac{\delta \tilde{p}_i(t)}{\delta q^j(t')} = - \frac{\partial^2 \Phi}{\partial q^i(t) \partial q^j(t)} \, \delta(t-t') \, , \qquad r=i+N, \; s=j \, , \tag{4.65c}$$

$$\langle r,t|\Omega|s,t'\rangle = \frac{\delta \tilde{p}_i(t)}{\delta p_j(t')} = \delta_j^{\;i} \, \delta(t-t') - \frac{\partial^2 \Phi}{\partial q^i(t) \partial p_j(t)} \, \delta(t-t') \, , \qquad r=i+N, \; s=j+N \, . \tag{4.65d}$$

Therefore, $\|\Omega\|$ is of the following form

$$\|\Omega\| = \|I\| \; + \; \|A\| \, , \tag{4.66}$$

where $\|I\|$ is the identity matrix while

$$\|A\| = \begin{bmatrix} \left[\frac{\partial^2\Phi}{\partial p_i(t)\partial q^j(t)}\,\delta(t-t')\right] & \left[\frac{\partial^2\Phi}{\partial p_i(t)\partial p_j(t)}\,\delta(t-t')\right] \\ \left[-\frac{\partial^2\Phi}{\partial q^i(t)\partial q^j(t)}\,\delta(t-t')\right] & \left[-\frac{\partial^2\Phi}{\partial q^i(t)\partial p_j(t)}\,\delta(t-t')\right] \end{bmatrix}. \tag{4.67}$$

As far as the discrete indices are concerned each sub-matrix in Eq.(4.67) is $N \times N$. On the other hand, $\|A\|$ is *diagonal* in the continuous indices. Next, we shall invoke the well-known relationship [4]

$$\begin{aligned} \Omega \equiv \det\|\Omega\| &= e^{\ln\det\|\Omega\|} = e^{\ln\det(\|I\|+\|A\|)} \\ &= 1 + \mathrm{tr}\|A\| + \mathcal{O}(\epsilon^2)\,. \end{aligned} \tag{4.68}$$

Now, from Eq.(4.67) it follows that

$$\begin{aligned} \mathrm{tr}\|A\| &= \int_{-\infty}^{+\infty} dt \sum_{r=1}^{2N} \langle r,t|A|r,t\rangle \\ &= \int_{-\infty}^{+\infty} dt \sum_{i=1}^{2N} \left[\frac{\partial^2\Phi}{\partial p_i(t)\partial q^i(t)} - \frac{\partial^2\Phi}{\partial q^i(t)\partial p_i(t)}\right]\delta(0) = 0 \end{aligned} \tag{4.69}$$

and, therefore,

$$[\mathcal{D}\tilde{q}]\,[\mathcal{D}\tilde{p}] = [\mathcal{D}q]\,[\mathcal{D}p]\,. \tag{4.70}$$

We turn next to analyzing $\delta[u^a(\tilde{q},\tilde{p})]$ and $\delta[t_a(\tilde{q},\tilde{p})]$. On the one hand,

$$u^a(\tilde{q},\tilde{p}) = u^a(q,p) + \bar{\delta}u^a(q,p)\,, \tag{4.71}$$

where

$$\bar{\delta}u^a(q,p) = [u^a(q,p), t_b(q,p)]_{PB}\,\epsilon^b\,. \tag{4.72}$$

On the other hand,

$$t_a(\tilde{q},\tilde{p}) = t_a(q,p) + \bar{\delta}t_a(q,p)\,, \tag{4.73}$$

where, on account of Eq.(4.2a),

$$\bar{\delta}t_a(q,p) = [t_a(q,p), t_b(q,p)]_{PB}\,\epsilon^b = t_c\,C^c_{ab}(q,p)\epsilon^b\,. \tag{4.74}$$

[4] Recall that according to Eqs.(4.67) and (4.62) all elements of the matrix $\|A\|$ are of order ϵ.

Hence,

$$t_a(\tilde{q},\tilde{p}) = t_c(q,p)\left(\delta^c_{\ a} + L^c_{\ a}(q,p)\right) , \tag{4.75}$$

where

$$L^c_{\ a}(q,p) \equiv C^c_{ab}(q,p)\epsilon^b . \tag{4.76}$$

Thus,

$$\prod_{a=1}^{M} \delta[u^a(\tilde{q},\tilde{p})] = \prod_{a=1}^{M} \delta[u^a(q,p) + \bar{\delta}u^a(q,p)] , \tag{4.77}$$

while (recall Eq.(B.97))

$$\begin{aligned}\prod_{a=1}^{M} \delta[t_a(\tilde{q},\tilde{p})] &= \prod_{a=1}^{M} \delta[t_c(q,p)\left(\delta^c_{\ a} + L^c_{\ a}(q,p)\right)] \\ &= \left[\det \|I + L\|\right]^{-1} \prod_{a=1}^{M} \delta[t_a(q,p)] . \end{aligned} \tag{4.78}$$

We look next for the modifications induced by the infinitesimal canonical transformation on the Faddeev-Popov determinant. To start with

$$\begin{aligned}&[u^a(\tilde{q},\tilde{p}), t_b(\tilde{q},\tilde{p})]_{PB} = [u^a(q,p) + \bar{\delta}u^a(q,p) \,,\, t_b(q,p) + \bar{\delta}t_b(q,p)]_{PB} \\ &= [u^a(q,p) + \bar{\delta}u^a(q,p) \,,\, t_c(q,p)]_{PB}\left(\delta^c_{\ b} + L^c_{\ b}(q,p)\right) \\ &+ t_c(q,p)[u^a(q,p) + \bar{\delta}u^a(q,p) \,,\, L^c_{\ b}(q,p)]_{PB} , \end{aligned} \tag{4.79}$$

implying that

$$\begin{aligned}&\left[\det \| \, [u^a(\tilde{q},\tilde{p}) \,,\, t_b(\tilde{q},\tilde{p})]_{PB} \, \| \right] \\ &= \Big[\det \| \, [u^a(q,p) + \bar{\delta}u^a(q,p), t_c(q,p)]_{PB}\left(\delta^c_{\ b} + L^c_{\ b}(q,p)\right) \\ &+ t_c(q,p)[u^a(q,p) + \bar{\delta}u^a(q,p) \,,\, L^c_{\ b}(q,p)]_{PB} \| \Big] , \end{aligned} \tag{4.80}$$

which along with Eq.(4.78) and

$$t_a(q,p)\,\delta[t_a(q,p)] = 0 \tag{4.81}$$

enables us to find

$$
\begin{aligned}
&\left(\prod_{a=1}^{M} \delta[t_a(\tilde{q},\tilde{p})]\right) \, [\det \| \, [u^a(\tilde{q},\tilde{p}) \, , \, t_b(\tilde{q},\tilde{p})]_{PB} \, \| \,] \\
&= \, [\det \|I + L\|]^{-1} \left(\prod_{a=1}^{M} \delta[t_a(q,p)]\right) \\
&\times \left[\det \| \, [u^a(q,p) \, + \, \bar{\delta}u^a(q,p), t_c(q,p)]_{PB} \, (\delta^c_{\ b} + L^c_{\ b}(q,p)) \, \|\right] \\
&= \left(\prod_{a=1}^{M} \delta[t_a(q,p)]\right) \left[\det \| [u^a(q,p) + \bar{\delta}u^a(q,p), t_b(q,p)]_{PB} \|\right] \, . \quad (4.82)
\end{aligned}
$$

The results so far obtained can be summarized as

$$
\begin{aligned}
&\left(\prod_{a=1}^{M} \delta[u^a(\tilde{q},\tilde{p})]\right) \left(\prod_{a=1}^{M} \delta[t_a(\tilde{q},\tilde{p})]\right) \, [\det \| \, [u^a(\tilde{q},\tilde{p}) \, , \, t_b(\tilde{q},\tilde{p})]_{PB} \, \| \,] \\
&= \left(\prod_{a=1}^{M} \delta[u^a(q,p) \, + \, \bar{\delta}u^a(q,p)]\right) \left(\prod_{a=1}^{M} \delta[t_a(q,p)]\right) \\
&\times \, \left[\det \| [u^a(q,p) + \bar{\delta}u^a(q,p), t_b(q,p)]_{PB} \|\right] \, , \quad (4.83)
\end{aligned}
$$

which arises from Eqs.(4.77) and (4.82).

What remains to be done is to determine the change in the action provoked by the infinitesimal canonical transformation under consideration. We concentrate first on the Hamiltonian

$$
\begin{aligned}
h(\tilde{q},\tilde{p}) &= h(q,p) + [h(q,p), t_a(q,p)]_{PB}\epsilon^a \\
&= h(q,p) + \sum_{b=1}^{M} t_b(q,p) B^b_a(q,p)\epsilon^a \, . \quad (4.84)
\end{aligned}
$$

Therefore,

$$
\begin{aligned}
&\left(\prod_{a=1}^{M} \delta[t_a(q,p)]\right) \exp\left\{-\frac{i}{\hbar}\int_{-\infty}^{+\infty} dt \, h(\tilde{q},\tilde{p})\right\} = \left(\prod_{a=1}^{M} \delta[t_a(q,p)]\right) \\
&\times \exp\left\{-\frac{i}{\hbar}\int_{-\infty}^{+\infty} dt \, [h(q,p) + \sum_{b=1}^{M} t_b(q,p) B^b_a(q,p)\epsilon^a]\right\} \\
&= \left(\prod_{a=1}^{M} \delta[t_a(q,p)]\right) \exp\left\{-\frac{i}{\hbar}\int_{-\infty}^{+\infty} dt \, h(q,p)\right\} \quad (4.85)
\end{aligned}
$$

where Eq.(4.81) was again invoked. As for the remaining term contributing to the action, it is not difficult to verify that

$$\int_{-\infty}^{+\infty} dt\, \tilde{p}_i \dot{\tilde{q}}^i = \int_{-\infty}^{+\infty} dt\, p_i \dot{q}^i + \left(p_i \frac{\partial \Phi}{\partial p_i} - \Phi \right) \Big|_{-\infty}^{+\infty}$$

$$= \int_{-\infty}^{+\infty} dt\, p_i \dot{q}^i + \epsilon^a \left(p_i \frac{\partial t_a}{\partial p_i} - t_a \right) \Big|_{-\infty}^{+\infty} . \tag{4.86}$$

When the constraints are first degree homogeneous functions of the momenta, the second term in the right hand side of Eq.(4.86) drops out and we end up with

$$\int_{-\infty}^{+\infty} dt\, \tilde{p}_i \dot{\tilde{q}}^i = \int_{-\infty}^{+\infty} dt\, p_i \dot{q}^i . \tag{4.87}$$

For constraints that are quadratic functions of the momenta, the surface term can be eliminated via a unitary transformation and Eq.(4.87) still holds. By collecting all results above we conclude that the effect of the infinitesimal canonical transformation on the path integral yielding $\mathbf{Z[J{=}0\,,K{=}0]}$ is that of replacing $u(q,p)$ by $u(q,p) + \bar{\delta} u(q,p)$. This secures the gauge independence of $\mathbf{Z[J{=}0\,,K{=}0]}$.

We close this subsection by mentioning that the connected Green functions generating functional for constrained systems is given by

$$e^{\frac{i}{\hbar} \tilde{W}_0 [J,K|+\infty,-\infty]} = \frac{\mathbf{Z[J\,,K]}}{\mathbf{Z[J = 0,K = 0]}} , \tag{4.88}$$

where

$$\mathbf{Z[J\,,K]}$$

$$= \int [\mathcal{D}q] \int [\mathcal{D}p] \left(\prod_{a=1}^{M} \delta[u^a]\, \delta[t_a] \right) [\det \| [u^a\,,\, t_b]_{PB} \|]$$

$$\times \exp \left\{ \frac{i}{\hbar} \int_{-\infty}^{+\infty} dt\, [p_j \dot{q}^j - h(q,p)] \right.$$

$$\left. + \frac{i}{\hbar} \int_{-\infty}^{+\infty} dt\, \left(q^i J_i + p_i K^i \right) \right\} . \tag{4.89}$$

4.3 Problem

Problem 4-1

Consider the Christ-Lee model [Christ and Lee (1980)]. It is defined by the Lagrangian

$$L = \frac{1}{2}\left(\dot{x}_1^2 + \dot{x}_2^2\right) - \left(x_1\dot{x}_2 - x_2\dot{x}_1\right)x_3 + \frac{1}{2}x_3^2\left(x_1^2 + x_2^2\right) - V(x_1^2 + x_2^2)\,, \tag{4.90}$$

where x_1, x_2 and x_3 are Cartesian coordinates while the overdot indicates differentiation with respect to time.

a) Find the Hamiltonian formulation of the dynamics of the Christ-Lee model [Costa and Girotti (1981)]. In particular, show that the constraint structure reduces to two first class constraints. Hence, this model is a gauge theory.

b) Find the local gauge transformations leaving the Lagrangian in Eq.(4.90) invariant. Hint: rewrite the Lagrangian in terms of cylindrical coordinates ($x_1 = r\cos\theta$, $x_2 = r\sin\theta$).

c) Show that the gauge conditions

$$b - c\arctan\frac{x_2}{x_1} \approx 0\,, \tag{4.91a}$$

$$x_3 \approx 0\,, \tag{4.91b}$$

where b and c are nonvanishing real constants, fix the gauge completely. Which is the value of the Faddeev-Popov determinant in this case? Construct explicitly the physical phase space.

d) Write down the functional integral $\mathbf{Z}[\mathbf{J}\,,\mathbf{K}]$ for the Christ-Lee model.

Chapter 5

Noncommutative systems

So far, we have been focusing on presenting and discussing the foundations of the functional formulation of quantum mechanics. We primarily aimed at untangling the subtleties of the formulation. All the models we analyzed served to this purpose.

We turn now to introduce the reader to a recent field of research: the functional quantization of a system involving noncommutative coordinates [Chaichian and Tureanu (2004); Gamboa and Rojas (2001, 2002); Horváthy and Plyushchay (2002); Girotti (2004); Bemfica and Girotti (2005)] [1].

It is a by-product of the theory of relativistic quantum fields set on a noncommutative space-time manifold, which were first considered in [Snyder (1947)]. It is worth mentioning that the idea that a noncommutative space-time manifold might cure the ultraviolet (UV) divergences of field theories was suggested by Heisenberg long before [Jackiw (2001)]. This discussion was abandoned due to the success of the renormalization theory so that its revival is rather recent and related to the string theory. Indeed, the noncommutative Yang-Mills theory arises as a limit of the string theory [Connes and Schwarz (1998)] and was extracted by Seiberg and Witten [Seiberg and Witten (1999)] starting from the open string in the presence of a magnetic field. More details on review articles [Rivelles (2000); Douglas and Nekrasov (2001); Szabo (2003); Gomes (2001); Girotti (2003)].

[1] There is a range of references concerning noncommutative quantum mechanics. We only mention here our papers of principle.

5.1 Classical-quantum transition for noncommutative systems

We shall be dealing with a quantum system whose dynamics is described by a self-adjoint Hamiltonian $H(Q,P)$ made up of the Cartesian coordinates $Q^l, l = 1,\ldots,N$ and their canonically conjugate momenta $P^j, j = 1,\ldots,N$. However, unlike the usual case, coordinates and momenta are supposed to obey the non-canonical equal-time commutation rules

$$[Q^l, Q^j] = -2i\hbar\theta^{lj}, \tag{5.1a}$$

$$[Q^l, P_j] = i\,\hbar\,\delta^l{}_j, \tag{5.1b}$$

$$[P_l, P_j] = 0. \tag{5.1c}$$

The distinctive feature, of course, is that the coordinates do not commute among themselves. The lack of noncommutativity of the coordinates is parametrized by the real antisymmetric $N \times N$ constant matrix $\|\theta\|$[2].

The classical counterpart of a quantum system involving non-commuting coordinates must be a constrained system. Indeed, the equal time algebra in Eq.(5.1a) could not have been abstracted from a Poisson bracket algebra, simply because the Poisson bracket of two coordinates vanishes. Moreover, the problem of finding a constrained system mapping onto the noncommutative theory specified in (5.1) was solved in [Deriglazov (2002)]. Its classical dynamics is described by the Lagrangian

$$L = v_j\,\dot{q}^j - h_0(q^j, v_j) + \dot{v}_j\,\theta^{jl}\,v_l\,, \tag{5.2}$$

where repeated indices are summed from 1 to N. The constraint structure of this system reduces to the primary second-class constraints $G_i \equiv p_i - v_i \approx 0\,, T^i \equiv \pi^i - \theta^{ij}v_j \approx 0\,$, where p_i (π^i) is the momentum canonically conjugate to the generalized coordinate q^i (v_i). As for the canonical Hamiltonian, we find

$$h(q,p) = h_0(q,p)\,. \tag{5.3}$$

We can check that the Faddeev-Popov matrix turns out to be unimodular and constant. Then, the computation of the Dirac brackets (DB) yields (recall section 4.1)

$$[q^j, q^k]_{DB} = -2\theta^{lk}, \tag{5.4a}$$

$$[q^j, p_k]_{DB} = \delta^j{}_k, \tag{5.4b}$$

$$[p_j, p_k]_{DB} = 0. \tag{5.4c}$$

[2]Observe that in the cgs system the dimension of θ^{lk} is $d[\theta] = \mathrm{g}^{-1}\,\mathrm{s}$.

We do not need to explicitly compute the DB's involving v's and/or π's since, by definition [Dirac (1964); Fradkin and Vilkovisky (1975); Costa and Simões (1985); Bemfica and Girotti (2008b)], within the DB algebra the constraints hold as strong identities. In fact, at this stage of the formulation it is possible to eliminate the variables v and π in favor of q and p. However, q and p may not be referred to as the physical phase space variables because their DB's differ from the corresponding Poisson brackets (PB). At this point, the construction of the physical phase space variables, (x, k), in terms of q and p is straightforward. Indeed, we can easily verify that

$$x^j \equiv q^j - \theta^{jl} p_l , \tag{5.5a}$$

$$k_j \equiv p_j \tag{5.5b}$$

and (5.4) lead to $[\xi^j, \xi^l]_{DB} = [\xi^j, \xi^l]_{PB}$, for ξ to be either x or k. All that remains to be done to erase any remaining trace of the constraints is to rewrite the Hamiltonian in (5.3) in terms of the physical variables, namely,

$$h(q,p) \equiv h\left(x^j + \theta^{jk} k_k , k_j\right) . \tag{5.6}$$

We can check that the Hamiltonian equations of motion for the physical variables possess the canonical form.

We turn next at quantizing the classical model described above. Within the operator framework, the quantization is implemented by first promoting x and k into the self-adjoint operators, X and K, respectively. The classical-quantum correspondence rule demands that they verify the equal-time commutator algebra

$$\left[X^l, X^j\right] = 0 , \tag{5.7a}$$

$$\left[X^l, K_j\right] = i\hbar \delta^l{}_j , \tag{5.7b}$$

$$[K_l, K_j] = 0 , \tag{5.7c}$$

while

$$H(X^j + \theta^{jl} K_l , K_j) = \frac{K_l K_l}{2M} + V(X^j + \theta^{jl} K_l) . \tag{5.8}$$

Hence, the quantum dynamics of a system involving noncommutative coordinates can be rephrased in terms of commutative ones; the price to be paid is a *non-local modification of the interaction.*

Then, in the Schrödinger picture, the development of the system in time is controlled by the wave equation

$$-\frac{\hbar^2}{2M}\nabla^2_{\tilde{x}}\,\psi(x,t) + V(x) * \psi(x,t) = i\hbar\frac{\partial\,\psi(x,t)}{\partial t}\,, \tag{5.9}$$

where

$$\begin{aligned} V(x) * \psi(x,t) &\equiv V\left(x^j - i\hbar\theta^{jk}\frac{\partial}{\partial x^k}\right)\psi(x,t) \\ &= V(x)\left[\exp\left(-i\hbar\,\frac{\overleftarrow{\partial}}{\partial x^j}\,\theta^{jk}\,\frac{\overrightarrow{\partial}}{\partial x^k}\right)\right]\psi(x,t)\,, \end{aligned} \tag{5.10}$$

is the Grönewold-Moyal (GM) product [Menzinescu (2000); Grönewold (1946); Moyal (1949); Filk (1996)]. This product is isomorphic to the product of composite operators $V(Q)\,\Psi(Q,t)$ defined on the manifold $[Q_l, Q_k] = -2i\hbar\,\theta_{lk} \neq 0$. Hence, we are effectively implementing quantum mechanics in a noncommutative manifold [Gamboa and Rojas (2001); Biggati and Susskind (2000)].

5.2 Example: the noncommutative two-dimensional harmonic oscillator

Before carrying out the task of finding the functional formulation of the quantum dynamics of systems involving noncommutative coordinates, we shall analyze the outcomes from the operator approach for the specific case of the noncommutative two-dimensional harmonic oscillator. Our goal is to gain insight into the modifications induced by noncommutativity.

The quantum mechanics in a noncommutative plane has been considered in [Gamboa and Rojas (2002)]. We shall now restrict to study the noncommutative two-dimensional harmonic oscillator of mass M and frequency ω. According to Eq.(5.8), the dynamic of the system is therefore determined by the Hamiltonian operator (H)

$$\begin{aligned} H(X,K) &= \frac{1}{2M}\left[K_l K^l + M^2\omega^2\left(X_j + \theta\varepsilon_{jl}K^l\right)\left(X^j + \theta\varepsilon^{jl}K_l\right)\right] \\ &\equiv \left(1 + M^2\omega^2\theta^2\right) H_\theta\,, \end{aligned} \tag{5.11}$$

where

$$H_\theta = \frac{1}{2M}\left(K_l K^l + M^2\omega_\theta^2\,X_l\,X^l + 2\,\theta\,M^2\,\omega_\theta^2\,\varepsilon_j{}^l\,X^j K_l\right) \tag{5.12}$$

and

$$\omega_\theta^2 \equiv \frac{\omega^2}{(1 + M^2\omega^2\,\theta^2)} \,. \tag{5.13}$$

Here, repeated indices run from 1 to 2. Furthermore, the antisymmetric matrix $||\theta||$ can be written as

$$\theta_{jk} = \theta\,\varepsilon_{jk}\,, \tag{5.14}$$

where ε_{jk} is the two-dimensional (antisymmetric) Levi-Civita tensor, ($\varepsilon_{12} = +1$) verifying $\varepsilon_j{}^k \varepsilon_{kl} = -\delta_{jl}$, while θ is a constant.

It will prove convenient to introduce next the creation ($a_j^\dagger$) and annihilation operators (a_j) obeying the following commutator algebra

$$[a_j\,,\,a_k] = 0\,, \tag{5.15a}$$
$$\left[a_j^\dagger\,,\,a_k^\dagger\right] = 0\,, \tag{5.15b}$$
$$\left[a_j\,,\,a_k^\dagger\right] = \delta_{jk}\,, \tag{5.15c}$$

in terms of which X_j, K_j can be written, respectively, as

$$X_j = \frac{1}{\sqrt{2}}\left(\frac{\hbar}{M\omega_\theta}\right)^{\frac{1}{2}}\left(a_j^\dagger + a_j\right)\,, \tag{5.16a}$$
$$K_j = \frac{1}{\sqrt{2}}\,(\hbar M\omega_\theta)^{\frac{1}{2}}\left(a_j^\dagger - a_j\right)\,. \tag{5.16b}$$

After substituting (5.16) into (5.12) we obtain

$$H_\theta = \hbar\omega_\theta\,(N + I + 2\theta M\omega_\theta\,J_2)\,, \tag{5.17}$$

where

$$N \equiv a_j^\dagger\, a_j\,, \tag{5.18a}$$
$$J_2 \equiv \frac{1}{2\hbar}\,L = \frac{1}{2\hbar}\,\varepsilon_j{}^l\,X^j K_l = -\,\frac{i}{2}\,a_j^\dagger\varepsilon_{jk}a_k\,, \tag{5.18b}$$

L is the angular momentum operator and I is the identity operator. We emphasize that the system under analysis possesses, as well as its commutative counterpart, the $SU(2)$ symmetry. The corresponding generators (J_1,

J_2, J_3) and the Casimir operator (J^2) are

$$J_1 = \frac{1}{2}\left(a_2^\dagger a_1 + a_1^\dagger a_2\right), \tag{5.19a}$$

$$J_2 = -\frac{i}{2}\, a_j^\dagger \varepsilon_{jk} a_k = \frac{i}{2}\left(a_2^\dagger a_1 - a_1^\dagger a_2\right), \tag{5.19b}$$

$$J_3 = \frac{1}{2}\left(a_1^\dagger a_1 - a_2^\dagger a_2\right), \tag{5.19c}$$

$$J^2 = J_k J_k = \frac{N}{2}\left(\frac{N}{2} + 1\right). \tag{5.19d}$$

Indeed, it is straightforward to verify that (5.15) implies both $[J_k, J_l] = i\varepsilon_{klm} J_m$ and $[J^2, J_k] = 0$, as required[3].

We shall denote by $|j, m\rangle$ the common eigenvectors of J^2 and J_2. They solve the following eigenvalue problem

$$J^2 |j, m\rangle = j(j+1) |j, m\rangle, \tag{5.20a}$$

$$J_2 |j, m\rangle = m |j, m\rangle, \tag{5.20b}$$

where, as we know, the eigenvalues j and m fulfill

$$j = 0, 1/2, 1, 3/2, \ldots, \tag{5.21a}$$

$$-j \leq m \leq j. \tag{5.21b}$$

Since Eqs.(5.19d) along with $[J^2, J_k] = 0$ imply that $[N, J_k] = 0$ we conclude that $\{|j, m\rangle\}$ are also eigenvectors of N and, therefore,

$$N |j, m\rangle = n |j, m\rangle. \tag{5.22}$$

The relation linking the eigenvalues j and n follows at once from Eqs.(5.19d), (5.20a) and (5.22). It reads

$$n = 2j. \tag{5.23}$$

Correspondingly, it follows from (5.17) that the energy eigenvalue problem can be cast as

$$H_\theta |j, m\rangle = \hbar\omega_\theta \left(n + 1 + 2\theta\, m\, M\, \omega_\theta\right) |j, m\rangle. \tag{5.24}$$

---

[3] ε_{klm} is the three-dimensional fully antisymmetric Levi Civita tensor with $\varepsilon^{123} = 1$.

In the commutative case ($\theta = 0$) the degeneracy of the nth energy level is $2j + 1 = n + 1$. We notice that *the degeneracy is lifted by the non-commutativity.*

The construction of $\{|j, m\rangle\}$ by acting with creation operators on a certain vacuum state goes as follows. We start by introducing new creation ($A_{\pm}^{\dagger}$) and annihilation ($A_{\pm}$) operators defined as

$$A_{\pm} \equiv \frac{1}{\sqrt{2}} \left(a_1 \mp i a_2\right) , \tag{5.25a}$$

$$A_{\pm}^{\dagger} \equiv \frac{1}{\sqrt{2}} \left(a_1^{\dagger} \pm i a_2^{\dagger}\right) , \tag{5.25b}$$

which fulfill the commutator algebra

$$\left[A_{\alpha} \, , \, A_{\beta}\right] = 0 \, , \tag{5.26a}$$

$$\left[A_{\alpha}^{\dagger} \, , \, A_{\beta}^{\dagger}\right] = 0 \, , \tag{5.26b}$$

$$\left[A_{\alpha} \, , \, A_{\beta}^{\dagger}\right] = \delta_{\alpha\beta} \, , \tag{5.26c}$$

where α and β are $+$ or $-$. Then, it turns out that[4]

$$|n_+, n_-\rangle = \frac{1}{\sqrt{n_+!}\sqrt{n_-!}} \left(A_+^{\dagger}\right)^{n_+} \left(A_-^{\dagger}\right)^{n_-} |0, 0\rangle \, , \tag{5.27}$$

are, for $n_{\pm}$ semi-positive definite integers and $A_{\pm}|0, 0\rangle = 0$, a complete and normalizable set of common eigenstates of the Hermitean operators

$$N_+ \equiv A_+^{\dagger} A_+ \, , \tag{5.28a}$$

$$N_- \equiv A_-^{\dagger} A_- \, . \tag{5.28b}$$

Since, by construction, $[N_+, N_-] = 0$ and, furthermore,

$$N = N_+ + N_- \, , \tag{5.29a}$$

$$J_2 = \frac{1}{2} \left(N_+ - N_-\right) , \tag{5.29b}$$

we conclude that the common eigenstates of energy and angular momentum can also be denoted by $|n_+, n_-\rangle$. The relationships among the labels follow from (5.23) and (5.29)

$$2j = n_+ + n_- \, , \tag{5.30a}$$

$$2m = n_+ - n_- \, . \tag{5.30b}$$

[4]See, for instance, [Messiah (1966)].

Of course, we may write [Gamboa and Rojas (2002)] instead of (5.27)

$$|j,m\rangle = \frac{1}{\sqrt{(j+m)!}\sqrt{(j-m)!}} \left(A_+^\dagger\right)^{(j+m)} \left(A_-^\dagger\right)^{(j-m)} |0,0\rangle \,. \tag{5.31}$$

5.3 Functional formulation of the quantum dynamics for noncommutative systems

In section 5.1, we saw that a noncommutative system maps onto a constrained system exhibiting second class constraints. The functional quantization of systems with constraints presented in section 4.2 is sufficient for this section.

Indeed, the starting point is the expression for the functional $\mathbf{Z}[\mathbf{J}\,,\mathbf{S}]$, given in Eq.(4.89) which, when written in terms of the phase space variables we used in section 5.1, is found to read

$$\mathbf{Z}[\mathbf{J}\,,\mathbf{S}] = \int [\mathcal{D}q] \int [\mathcal{D}v] \int [\mathcal{D}p] \int [\mathcal{D}\pi] \left\{ \prod_{j=1}^{N} \delta[p_j - v_j] \right\}$$
$$\times \left\{ \prod_{j=1}^{N} \delta[\pi^j - \theta^{jk} v_k] \right\} \exp\left\{ \frac{i}{\hbar} \int_{t_{in}}^{t_f} dt \left[p_j\, \dot{q}^j \,+\, \pi^j\, \dot{v}_j \,-\, h(q,p) \right.\right.$$
$$\left.\left. +\, q^j\, J_j \,+\, p_j\, S^j \right] \right\}. \tag{5.32}$$

We have just ignored the Faddeev-Popov determinant because it is an irrelevant constant for the model under scrutiny. After performing the functional integrals on π and v we end up with

$$\mathbf{Z}[\mathbf{J}\,,\mathbf{S}] = \int [\mathcal{D}q] \int [\mathcal{D}p]$$
$$\times \exp\left\{ \frac{i}{\hbar} \int_{t_i}^{t_f} dt \left[p_j \dot{q}^j \,-\, p_j \theta^{jk} \dot{p}_k - h(q,p) + q^j J_j + p_j\, S^j \right] \right\}. \tag{5.33}$$

So far, we have succeeded in eliminating all the redundant degrees of freedom and, therefore, in expressing $\mathbf{Z}[\mathbf{J}\ ,\ \mathbf{S}]$ as a phase space path integral over independent variables. Yet this is not the end of the story since, as mentioned, q and p are not canonical phase space variables. However, an expression for $\mathbf{Z}[\mathbf{J}\ ,\ \mathbf{S}]$ written as a phase space path integral over independent canonical variables $(x,\ k)$ was shown to exist (see section 4.2).

Presently, we can find such expression by performing the non-canonical transformation in Eq.(5.5) which, in turn, allows us to cast Eq.(5.33) as

$$\mathbf{Z}[\mathbf{J}\ ,\ \mathbf{S}] = \int [\mathcal{D}x] \int [\mathcal{D}k]$$
$$\times \exp\left\{\frac{i}{\hbar}\int_{t_i}^{t_f} dt\, \left[k_j \dot{x}^j \ -\ h(x^j + \theta^{jl} k_l, k_j) + x^j V_j + k_j\, U^j\right]\right\}, \quad (5.34)$$

where [5]

$$V_j \equiv J_j\,, \quad (5.35a)$$
$$U^j \equiv S^j \ -\ \theta^{j\,k}\, J_k\,. \quad (5.35b)$$

We now bring again into play the noncommutative two-dimensional harmonic oscillator as a testing ground for the functional formulation of noncommutative systems. The classical counterpart of the Hamiltonian operator in Eq.(5.11) reads

$$h(x^j + \theta^{jl} k_l, k_j) \ =\ \frac{k_j k_j}{2M} \ +\ \frac{M\omega^2}{2}\left(x^j x^j + 2x^i \theta^{ij} k_j + \theta^{ij}\theta^{il} k_j k_l\right)\ . \quad (5.36)$$

We substitute this into Eq.(5.34). By expressing the result in terms of the propagator [6] we find, at the limit of vanishing external sources,

$$K(x_f, t_f; x_i, t_i) \ =\ \int [\mathcal{D}x]\ [\mathcal{D}k]\ \exp\left\{\frac{i}{\hbar}\int_{t_{in}}^{t_f} dt\, \left[k_j\, \dot{x}^j \ -\ h(x^j + \theta^{jl} k_l, k_j)\right]\right\}\ . \quad (5.37)$$

After carrying out the momentum integrals we arrive at

$$K(x_f, t_f; x_i, t_i) \ =\ \int \mathcal{D}x\, e^{\frac{i}{\hbar}\, S_{eff}[x]}\,, \quad (5.38)$$

where

$$S_{eff}[x] \ =\ \int_{t_i}^{t_f} dt\, L_{eff}(x(t), \dot{x}(t)) \quad (5.39)$$

[5]We remind the reader that the functional integral on x is carried out over the collection of functions verifying the boundary conditions $x(t_i) = x_i, x(t_f) = x_f$.

[6]A detailed discussion of the relationship linking the propagator with the generating functional of Green functions is given in Chapter 3.

denotes the effective action,

$$L_{eff}(x(t),\dot{x}(t)) = \frac{1}{2} M_\theta \, \dot{x}^j \dot{x}_j - M_\theta \, M \, \omega^2 \, \theta \, x^i \, \epsilon_{ij} \, \dot{x}^j - \frac{1}{2} M_\theta \, \omega^2 \, x^j \, x_j \,, \tag{5.40}$$

is the effective Lagrangian and

$$M_\theta \equiv \frac{M}{1 + M^2 \omega^2 \theta^2} \,. \tag{5.41}$$

Clearly, the effective Lagrangian in Eq.(5.40) is quadratic, as well as its commutative counterpart. Therefore, the path integral in Eq.(5.38) can be exactly computed. It yields

$$K(x_f, t_f; x_i, t_i) = \mathcal{N} \, e^{\frac{i}{\hbar} S_{eff}[x_{cl}]} \,, \tag{5.42}$$

where $\mathcal{N}$ is a normalization constant while x_{cl} are the solutions of the Lagrange equations of motion

$$\ddot{x}^1 + 2 \, \theta \, M \, \omega^2 \, \dot{x}^2 + \omega^2 \, x^1 = 0 \,, \tag{5.43a}$$

$$\ddot{x}^2 - 2 \, \theta \, M \, \omega^2 \, \dot{x}^1 + \omega^2 \, x^2 = 0 \,, \tag{5.43b}$$

deriving from the effective Lagrangian (5.40). We can convince ourselves that Eqs.(5.39), (5.40) and (5.43) lead to

$$S_{eff}[x_{cl}] = \frac{1}{2} M_\theta \left[x^j_{cl}(t_f) \, \dot{x}^j_{cl}(t_f) - x^j_{cl}(t_i) \, \dot{x}^j_{cl}(t_i) \right] . \tag{5.44}$$

When the solutions for x_{cl} are substituted back into the right hand side of Eq.(5.44) we obtain (see Problem 8)

$$\begin{aligned} S_{eff}[x_{cl}] = \; & \frac{M_\theta}{2} \frac{\omega\sqrt{\kappa}}{\sin\left[\omega\sqrt{\kappa}T\right]} \left\{ \left(x^j_f x^j_f + x^j_i x^j_i \right) \cos\left[\omega\sqrt{\kappa}T\right] \right. \\ & - 2 x^j_i x^j_f \cos\left[M\theta\omega^2 T\right] \\ & \left. + \, 2\epsilon^{jk} x^j_f x^k_i \sin\left[M\theta\omega^2 T\right] \right\} , \end{aligned} \tag{5.45}$$

where, as always, $T = t_f - t_i$ and

$$\kappa \equiv 1 + M^2 \theta^2 \omega^2 \,. \tag{5.46}$$

5.4 Noncommutative systems and the time slicing definition of the path integral

We shall next investigate the functional formulation of the quantum dynamics of noncommutative systems when the phase space path integral is defined through the time slicing procedure. The central issue is, of course, the uniqueness of the classical-quantum transition for this kind of systems. For this purpose, we are using the noncommutative two-dimensional harmonic oscillator.

As seen in Chapter 1, the dependence of the results on the discretization (α-dependence) may arise from two different sources. One of them is the α-dependence of the GWT ($h_{\theta_\alpha}(x,k)$) of the Hamiltonian operator in Eq.(5.11)

$$H(X,K) = \frac{1}{2M}\left[K_l K^l + M^2\omega^2\left(X_j + \theta\varepsilon_{jl}K^l\right)\left(X^j + \theta\varepsilon^{jl}K_l\right)\right]. \quad (5.47)$$

The other one originates from the replacement

$$x_a^j \longrightarrow \left(\frac{1}{2} - \alpha\right) x_a^j + \left(\frac{1}{2} + \alpha\right) x_{a+1}^j\,, \quad (5.48)$$

within the argument of $h_{\theta_\alpha}(x,k)$ [7].

The Hamiltonian operator in Eq.(5.47) can be rearranged as follows

$$\begin{aligned} H(X,K) = &\left(\frac{1}{2M} + \frac{M\omega^2\theta^2}{2}\right) K^j K_j + \frac{M\omega^2}{2} X^j X_j \\ &+ \frac{M\omega^2}{2}\,\theta\left(X_j \epsilon^{jl} K_l + K_l \epsilon^{jl} X_j\right). \end{aligned} \quad (5.49)$$

The first two terms in the right hand side of this last equation do not involve products of noncommuting operators. Therefore, in accordance with our developments in Chapter 1, the corresponding GWT's do not depend on α and read

$$\left(\frac{1}{2M} + \frac{M\omega^2\theta^2}{2}\right) K^j K_j \xrightarrow{\alpha} \left(\frac{1}{2M} + \frac{M\omega^2\theta^2}{2}\right) k^j k_j\,, \quad (5.50)$$

$$\frac{M\omega^2}{2} X^j X_j \xrightarrow{\alpha} \frac{M\omega^2}{2} x^j x_j\,. \quad (5.51)$$

[7] We shall be dealing here with two different kinds of discrete indices. On the one hand, lower case letters from the beginning of the Latin alphabet label time slices and run from 0 to m. On the other hand, lower case letters from the middle of Latin alphabet label Cartesian components of coordinates (x) and momenta (k) and run from 1 to 2.

As for the third term, *it does contain* products of noncommuting operator which signalizes, at least in principle, the presence of α dependent terms in its GWT. However, by starting from Eq.(1.52) we find

$$X_j \epsilon^{jl} K_l + K_l \epsilon^{jl} X_j \xrightarrow{\alpha} 2\epsilon^{jl} x_j k_l - 2\left(\frac{1}{2} - \alpha\right) \frac{\hbar}{i} \epsilon^{jl} \delta_{jl}$$
$$= 2\epsilon^{jl} x_j k_l \,. \tag{5.52}$$

Thus, the *would be* α-dependence in $h_{\theta_\alpha}(x,k)$ is washed out by the antisymmetry of the tensor ϵ^{jl} associated with the noncommutativity [8].

To summarize:

$$H(X,K) \xrightarrow{\alpha} h_\theta(x,k)\,, \tag{5.53}$$

where

$$h_\theta(x,k) = \left(\frac{1}{2M} + \frac{M\omega^2\theta^2}{2}\right) k^j k_j + \frac{M\omega^2}{2} x^j x_j$$
$$+ M\omega^2\, \theta x_j \epsilon^{jl} k_l\,, \tag{5.54}$$

which, after the replacement specified in Eq.(5.48), goes into

$$h_\theta(x(\alpha),k) = \left(\frac{1}{2M} + \frac{M\omega^2\theta^2}{2}\right) k^j k_j + \frac{M\omega^2}{2} x^j(\alpha) x_j(\alpha)$$
$$+ M\omega^2\, \theta x_j(\alpha) \epsilon^{jl} k_l\,. \tag{5.55}$$

We have reached the point where it becomes possible to write down the expression for the propagator of the noncommutative two-dimensional harmonic oscillator regarding the phase space path integral defined through the time slicing procedure. The starting point is the expression in Eq.(1.128) specialized for the present situation, i.e.,

$$K(x_f, t_f; x_i, t_i) = \lim_{m\to\infty} (2\pi\hbar)^{-2(m+1)} \int_{-\infty}^{+\infty} \left(\prod_{a=1}^{m} d^2 x_a\right) \int_{-\infty}^{+\infty} \left(\prod_{a=0}^{m} d^2 k_a\right)$$
$$\times\, e^{\frac{i}{\hbar}\sum_{a=0}^{m} \epsilon \left[k_a^j \frac{(x_{a+1}^j - x_a^j)}{\epsilon} - h_\theta(x(\alpha),k)\right]}\,. \tag{5.56}$$

[8] This result also applies for cases other than the noncommutative two-dimensional harmonic oscillator [Bemfica and Girotti (2008a); Bemfica (2009)].

The momentum integrals in Eq.(5.56) can be carried out with the help of Eq.(A.10) and, after algebraic rearrangements, we arrive at

$$
\begin{aligned}
K(x_f, t_f; x_i, t_i) &= \lim_{m\to\infty}\left(\frac{M_\theta}{2\pi\hbar i\epsilon}\right)^{(m+1)}\int_{-\infty}^{+\infty}\left(\prod_{a=1}^{m} d^2x_a\right)\\
&\times \exp\left\{\frac{i}{\hbar}\sum_{a=0}^{m}\epsilon\left[\frac{M_\theta}{2}\frac{\left(x^j_{a+1}-x^j_a\right)}{\epsilon}\frac{\left(x^j_{a+1}-x^j_a\right)}{\epsilon}\right.\right.\\
&\left.\left.-MM_\theta\omega^2\theta x^l_a(\alpha)\epsilon_{lj}\frac{\left(x^j_{a+1}-x^j_a\right)}{\epsilon} - M_\theta\frac{\omega^2}{2}x^j_a(\alpha)x^j_a(\alpha)\right]\right\}\\
&= \lim_{m\to\infty}\left(\frac{M_\theta}{2\pi\hbar i\epsilon}\right)^{(m+1)}\int_{-\infty}^{+\infty}\left(\prod_{a=1}^{m} d^2x_a\right)\\
&\times \exp\left\{\frac{i}{\hbar}\sum_{a=0}^{m}\frac{M_\theta}{2\epsilon}\left[\left(x^j_{a+1}-x^j_a\right)\left(x^j_{a+1}-x^j_a\right)\right.\right.\\
&\left.\left.-2\epsilon M\omega^2\theta x^l_a(\alpha)\epsilon_{lj}\left(x^j_{a+1}-x^j_a\right)-\omega^2\epsilon^2x^j_a(\alpha)x^j_a(\alpha)\right]\right\},
\end{aligned}
\tag{5.57}
$$

where M_θ and κ were defined in Eqs.(5.41) and (5.46), respectively.

We now focus on the the second term in the exponent in Eq.(5.57). To start with, it couples the coordinate components x^1 and x^2. Furthermore, it appears to exhibit bilinear terms of the form $\epsilon\alpha$ which might jeopardize the uniqueness of the classical-quantum transition. Nevertheless, this last difficulty is only apparent because

$$
\begin{aligned}
x^j_a(\alpha)\epsilon_{jl}\left(x^l_{a+1}-x^l_a\right) &= \left[\left(\frac{1}{2}+\alpha\right)x^j_{a+1}+\left(\frac{1}{2}-\alpha\right)x^j_a\right]\epsilon_{jl}\left(x^l_{a+1}-x^l_a\right)\\
&= -x^j_{a+1}\epsilon_{jl}x^l_a\,,
\end{aligned}
\tag{5.58}
$$

which does not contain α-dependent terms. It is worth mentioning that to arrive at this result we used

$$
\left(\frac{1}{2}+\alpha\right)x^j_{a+1}\epsilon_{jl}x^l_{a+1} = \left(\frac{1}{2}-\alpha\right)x^j_a\epsilon_{jl}x^l_a = 0\,, \tag{5.59}
$$

which are consequences of the antisymmetry of the tensor ϵ_{jl}. Once again, the antisymmetry of the tensor ϵ_{jl} plays a key role in the elimination of a dangerous α-dependence. The remaining α-dependent terms cancel out among themselves.

In short,

$$\sum_{a=0}^{m} \frac{M_\theta}{2\epsilon} \Big[\left(x^j_{a+1} - x^j_a\right)\left(x^j_{a+1} - x^j_a\right) \\ -2\epsilon M\omega^2\theta x^l_a(\alpha)\epsilon_{lj}\left(x^j_{a+1} - x^j_a\right) - \omega^2\epsilon^2 x^j_a(\alpha)x^j_a(\alpha)\Big] \\ = R_{\theta_{jl}}(\alpha)x^j_{a+1}x^l_{a+1} + 2G_{\theta_{jl}}(\alpha)x^j_{a+1}x^l_a + H_{\theta_{jl}}(\alpha)x^j_a x^l_{a+1} \tag{5.60}$$

with

$$R_{\theta_{jl}}(\alpha) = \frac{M_\theta}{2\epsilon}\left[1 - \epsilon^2\omega^2\left(\frac{1}{2} + \alpha\right)^2\right]\delta_{jl}\,, \tag{5.61a}$$

$$G_{\theta_{jl}}(\alpha) = \frac{M_\theta}{2\epsilon}\left[-\delta_{jl} + \epsilon M\omega^2\theta\epsilon_{jl} - \epsilon^2\omega^2\left(\frac{1}{4} - \alpha^2\right)\delta_{jl}\right]\,, \tag{5.61b}$$

$$H_{\theta_{jl}}(\alpha) = \frac{M_\theta}{2\epsilon}\left[1 - \epsilon^2\omega^2\left(\frac{1}{2} - \alpha\right)^2\right]\delta_{jl}\,. \tag{5.61c}$$

Clearly, at the limit $\theta \to 0$ ($M_\theta \to M$) we are left with two uncoupled one-dimensional harmonic oscillators (compare with Eq.(1.139)).

To a certain extent, what comes next parallels the developments after Eq.(1.139). We leave it as an exercise for the reader to complete the evaluation of the propagator for the noncommutative two-dimensional harmonic oscillator and to confirm that the final result turns out to be independent of α and agrees with the expression given at Eq.(5.42) (Problem 10).

5.5 Problems

Problem 5-1

Verify that:

a) the constraint structure following from the Lagrangian in Eq.(5.2) is that mentioned in the paragraph following this equation,

b) the corresponding Faddeev-Popov matrix is constant and the modulus of its determinant turns out to be 1 (unimodular),

c) the expressions in Eq.(5.4), for the corresponding DB's, are correct.

Problem 5-2

a) Supply a proof for the equality in Eq.(5.10). Prove that if $V(x)$ is a polynomial, the series in the right hand side of Eq.(5.10) terminates.

b) Let $\phi_1(x)$ and $\phi_2(x)$ be two infinitely differentiable real functions. Demonstrate that, up to surface terms,

$$\int dx\, \phi_1(x) \star \phi_2(x) = \int dx\, \phi_1(x)\phi_2(x)\,. \tag{5.62}$$

In words, under the integral sign the GM product of two functions reduces to the ordinary product.

c) The integral of the GM product,

$$\int dx\, \phi_1(x) \star \phi_2(x) \star \cdots \star \phi_n(x)\,, \tag{5.63}$$

turns out to be invariant under cyclic permutations.

d) Assume that the GM product is associative. Then, show that

$$\int dx\, \phi_1(x) \star \phi_2(x) \star \phi_3(x) = \int dx\, \phi_3(x)\left[\phi_1(x) \star \phi_2(x)\right]\,. \tag{5.64}$$

Problem 5-3

a) By combining results obtained in the present section, show that the energy eigenvalues ($E_{n_+n_-}$) of the two-dimensional noncommuting harmonic oscillator can be cast as

$$\begin{aligned} E_{n_+n_-} &= \sqrt{1 + M^2\theta^2\omega_\theta^2}\,\hbar\omega \\ &\times \left[1 + (1 + M\theta\omega_\theta)\, n_+ + (1 - M\theta\omega_\theta)\, n_-\right]\,. \end{aligned} \tag{5.65}$$

b) Consider this system in thermodynamical equilibrium with a heat reservoir at temperature T. By definition, the corresponding partition function is given by

$$Z_\theta(\mu) = Tre^{-\mu H}\,, \tag{5.66}$$

where Tr designates the operation of taking the trace. Furthermore, $\mu = 1/kT$ and k is the Boltzmann constant. Demonstrate that

$$Z_\theta(\mu) = \frac{1}{4\sinh\left(\frac{y_\theta s_\theta^+}{2}\right)\sinh\left(\frac{y_\theta s_\theta^-}{2}\right)}, \tag{5.67}$$

where

$$y_\theta \equiv \mu\hbar\omega_\theta\left(1 + M^2\omega^2\theta^2\right), \tag{5.68a}$$
$$s_\theta^\pm \equiv 1 \pm M\omega_\theta\theta. \tag{5.68b}$$

Problem 5-4

a) The mean energy ($\langle E_\theta\rangle$) is defined as

$$\langle E_\theta\rangle \equiv -\left.\frac{\partial\left(\ln Z_\theta(\mu)\right)}{\partial\mu}\right|_{\mu=\frac{1}{kT}}. \tag{5.69}$$

For the quantized noncommutative two-dimensional harmonic oscillator, show that

$$\langle E_\theta\rangle = \frac{\hbar\omega_\theta}{2}\left[\frac{1}{s_\theta^-}\coth\left(\frac{y_\theta s_\theta^+}{2}\right) + \frac{1}{s_\theta^+}\coth\left(\frac{y_\theta s_\theta^-}{2}\right)\right]. \tag{5.70}$$

b) Verify that in the commutative limit the right hand side in Eq.(5.70) reduces, as it must, to the Planck's formula,

$$\langle E_{\theta=0}\rangle = \hbar\omega\coth\left(\frac{\hbar\omega}{2kT}\right). \tag{5.71}$$

Problem 5-5

a) Cast the right hand side in Eq.(5.70) as

$$\langle E_\theta\rangle = \hbar\omega\left\{\sqrt{1+z^2} + \frac{\sqrt{1+z^2}+z}{\exp\left[\frac{\hbar\omega}{kT}\left(\sqrt{1+z^2}+z\right)\right]-1} + \frac{\sqrt{1+z^2}-z}{\exp\left[\frac{\hbar\omega}{kT}\left(\sqrt{1+z^2}-z\right)\right]-1}\right\}, \tag{5.72}$$

where z is the dimensionless variable

$$z = M\omega\theta \tag{5.73}$$

b) Assume that ω and z are fixed, while T varies. Then, prove that

$$\lim_{\hbar\omega \ll kT} \langle E_\theta \rangle \longrightarrow 2\,k\,T\,, \tag{5.74}$$

implying that the noncommutativity does not alter the high temperature limit. Show that, at the other end of the scale temperature, we find

$$\lim_{\hbar\omega \gg kT} \langle E_\theta \rangle \longrightarrow \hbar\omega\,\sqrt{1+z^2}\,, \tag{5.75}$$

which, as expected, coincides with the energy eigenvalue in Eq.(5.65) for $n_+ = n_- = 0$.

c) In the sequel, the interested reader is invited to test the limits of infinite noncommutativity, $\theta \to \pm\infty$, while keeping ω, M and T fixed. You should demonstrate that

$$\lim_{|\theta| \to +\infty} \langle E_\theta \rangle \longrightarrow \hbar\omega \left(M\omega|\theta| + \frac{kT}{\hbar\omega} \right) \longrightarrow \hbar\omega^2 M|\theta|\,. \tag{5.76}$$

Compare and discuss this result with that obtained for the high temperature limit in the commutative case. Which quantity appears to play the role of the *high temperature*?

Problem 5-6

The second term in the right hand side of Eq.(5.40) describes the interaction of an electrically charged particle (charge e) with a constant magnetic field B. Find the components of the magnetic vector potential $\vec{A}$ and show that the corresponding magnetic field turns out to be given by

$$B = \vec{\nabla} \times \vec{A} = -2\frac{M_\theta M\omega^2\theta c}{e} = \text{constant}\,, \tag{5.77}$$

where c is the speed of light in vacuum. In this sense, the noncommutative two-dimensional harmonic oscillator maps onto the Landau problem [Landau and Lifshitz (1958)].

Problem 5-7

By following the systematics in section 3.3, verify the commutation relations in Eq.(5.1) by starting from Eq.(5.34). *Hint*: Since the equal time commutation relations in Eq.(5.1) are not modified by the interaction the calculations may be simplified by choosing $h(x^j + \theta^{jl} k_l, k_j)$ equal to the free Hamiltonian, namely,

$$h(x^j + \theta^{jl} k_l, k_j) = \frac{1}{2M} k_j k_j \,. \tag{5.78}$$

Problem 5-8

a) Solve the system of coupled differential equations in Eq.(5.43).
b) Implement the steps leading to Eq.(5.45).
c) Analyze the commutative limit of Eq.(5.45). Compare the result with the outcome for the propagator of the ordinary harmonic oscillator (see Eq.(1.176).

Problem 5-9

Show that for any operator function $V(X, K)$, where X and K enter the argument of V *only* through the combination $X^j + \theta^{jl} K_l$ and θ^{jl} is a constant antisymmetric tensor and that it follows that

$$V(X, K) \xrightarrow{\alpha} V_\theta(x^j + \theta^{jl} k_l)\, e^{-\frac{i}{\hbar}\frac{1}{\left(\frac{1}{2}-\alpha\right)} k_j \theta^{jl} k_l} = V_\theta(x^j + \theta^{jl} k_l) \,. \tag{5.79}$$

As mentioned before, the antisymmetry of the tensor θ^{jl}, parametrizing the noncommutativity, kills the α-dependence in the right hand side of Eq.(5.79).

Problem 5-10

Complete the evaluation of the propagator for the noncommutative two-dimensional harmonic oscillator and show that the final result turns out to be independent of α and agrees with the expression given at Eq.(5.42) with $S_{eff}[x_{cl}]$ given at Eq.(5.45).

Appendix A

Some useful integrals

In this appendix we gathered a set of results about definite integrals which are needed for the developments in the text.

Let a be a real positive constant and $x \in \mathbb{R}_1$. Then,

$$\int_{-\infty}^{+\infty} dx\, e^{-\,ax^2} = \sqrt{\frac{\pi}{a}} \tag{A.1}$$

and also

$$\int_{-\infty}^{+\infty} \frac{dx}{\sqrt{2\pi}}\, e^{-\,\frac{1}{2}\,x^2} = 1\,, \tag{A.2}$$

$$\int_{-\infty}^{+\infty} dx\, e^{-\,a(x+x_0)^2} = \sqrt{\frac{\pi}{a}}\,. \tag{A.3}$$

As a by-product we have that

$$\int_{-\infty}^{+\infty} dx\, e^{-\,ax^2+bx} = e^{\frac{b^2}{4a}}\sqrt{\frac{\pi}{a}}\,. \tag{A.4}$$

However, the integrals belonging to the functional formulation of Quantum Mechanics are of the type

$$\int_{-\infty}^{+\infty} dx\, e^{i\,ax^2}\,, \tag{A.5}$$

which is not well defined due to the fact that its integrand oscillates with no limit at the endpoints of the interval. However, it is still possible to assign it a meaning by making use of a regularization procedure based on

$$\int_{-\infty}^{+\infty} dx\, e^{i\,ax^2} \equiv \lim_{\eta\to 0}\int_{-\infty}^{+\infty} dx\, e^{i\,ax^2-\eta x^2}\,, \tag{A.6}$$

where η is a positive definite constant. We can evaluate Eq.(A.6) by using Eq.(A.1)

$$\int_{-\infty}^{+\infty} dx\, e^{i\,ax^2-\eta x^2} = \sqrt{\frac{\pi}{\eta - ia}}\,. \tag{A.7}$$

Then,

$$\int_{-\infty}^{+\infty} dx\, e^{i\,ax^2} = \begin{cases} \sqrt{\frac{i\,\pi}{a}}\,, & \text{if} \quad a > 0 \\ \sqrt{\frac{\pi}{i|a|}}\,, & \text{if} \quad a < 0 \end{cases}. \tag{A.8}$$

One can easily verify that

$$\int_{-\infty}^{+\infty} dx\, e^{i\,a(x+x_0)^2} = \begin{cases} \sqrt{\frac{i\,\pi}{a}}\,, & \text{if} \quad a > 0 \\ \sqrt{\frac{\pi}{i|a|}}\,, & \text{if} \quad a < 0 \end{cases} \tag{A.9}$$

and

$$\int_{-\infty}^{+\infty} dx\, e^{i\,ax^2+ibx} = \begin{cases} e^{-i\,\frac{b^2}{4a}}\sqrt{\frac{i\,\pi}{a}}\,, & \text{if} \quad a > 0 \\ e^{+i\,\frac{b^2}{4|a|}}\sqrt{\frac{\pi}{i|a|}}\,, & \text{if} \quad a < 0 \end{cases} \tag{A.10}$$

follow from Eq.(A.8).

Let $|X\rangle$ be a vector and $\{|j\rangle, j = 1, \ldots, N\}$ an orthonormal basis in $\mathbb{R}_N$. Therefore, we can write

$$I = \sum_{j=1}^{N} |j\rangle\langle j|\,, \tag{A.11}$$

where I is the identity operator. We denote by A the self-adjoint operator whose domain is $\mathcal{D}_A = \mathbb{R}_N$. Furthermore,

$$x_j \equiv \langle j|X\rangle \tag{A.12}$$

and

$$a_{ij} = \langle i|A|j\rangle \tag{A.13}$$

are, respectively, the components of $|X\rangle$ and the elements of the matrix representative of the operator A in the just mentioned basis. We claim that

$$\int_{-\infty}^{+\infty}\left(\prod_{j=1}^{N} dx_j\right)\exp\left(\frac{i}{\hbar}\sum_{j,k=1}^{N} x_j a_{jk} x_k\right) = (i\pi\hbar)^{\frac{N}{2}}\,(\det A)^{-\frac{1}{2}}\,. \tag{A.14}$$

Before proceeding we have some remarks in order. The operator A defines the eigenvalue problem

$$A\,|\lambda^{(j)}\rangle = \lambda^{(j)}\,|\lambda^{(j)}\rangle\,. \tag{A.15}$$

Since we are dealing with a self-adjoint operator the eigenvalues $\{\lambda^{(j)},\, j = 1,\ldots,N\}$ are real while the eigenvectors $\{|\lambda^{(j)}\rangle,\, j = 1,\ldots,N\}$ corresponding to different eigenvalues are orthogonal. Let us suppose that there is no degeneracy, which implies that, after normalization, the set of eigenvectors provides another orthonormal basis, i.e.,

$$I = \sum_{j=1}^{N} |\lambda^{(j)}\rangle\langle\lambda^{(j)}|\,. \tag{A.16}$$

As for $\det A$, it is given by

$$\det A = \prod_{j=1}^{N} \lambda^{(j)}\,. \tag{A.17}$$

Hence, the expression in Eq.(A.14) only makes sense if the operator A does not possess zero modes. From here on, we shall be dealing only with operators whose eigenvalues are positive definite.

We now face the task of proving Eq.(A.14). The strategy consists in reformulating this integral by using the basis of the eigenvectors of the operator A. From Eqs.(A.12), (A.13) we obtain

$$\begin{aligned}\sum_{j,k=1}^{N} x_j a_{jk} x_k &= \sum_{j,k=1}^{N} \langle X|j\rangle\langle j|A|k\rangle\langle k|X\rangle \\ = \langle X|A|X\rangle &= \sum_{j,k=1}^{N} \langle X|\lambda^{(j)}\rangle\langle\lambda^{(j)}|A|\lambda^{(k)}\rangle\langle\lambda^{(k)}|X\rangle \\ = \sum_{j=1}^{N} \lambda^{(j)}\, x_j'^{\,2}\,, \end{aligned} \tag{A.18}$$

where

$$x'_j \equiv \langle \lambda^{(j)} | X \rangle \,. \tag{A.19}$$

The problem of transforming the integrand in Eq.(A.14) is solved. What remains to be done is to find the new integration measure. The operator

$$U \equiv \sum_{i=1}^{N} |i\rangle\langle \lambda^{(i)}| \tag{A.20}$$

implements the canonical transformation from the basis $\{|\lambda^{(i)}\rangle\}$ to the basis $\{|i\rangle\}$. Clearly, U is a unitary operator, a property also shared by its matrix representative $\|U\|$. However, since we are working in a real linear vector space, the Hermitian conjugate of this matrix coincides with its transpose $\|U\|^T$. Thus, the unitarity condition reduces to the orthogonality condition

$$\|U\|\,\|U\|^T = \|U\|^T\,\|U\| = I \Longrightarrow \left|\det \|U\|\right| = +1\,. \tag{A.21}$$

However, $\left|\det \|U\|\right|$ is the Jacobian of the transformation $x \rightarrow x'$ and, therefore,

$$\prod_{J=1}^{N} dx'_j = \prod_{J=1}^{N} dx_j\,. \tag{A.22}$$

By taking into account Eqs.(A.18), (A.22) and (A.8) we achieve

$$\begin{aligned}
&\int_{-\infty}^{+\infty} \left(\prod_{j=1}^{N} dx_j \right) \exp\left(\frac{i}{\hbar} \sum_{j,k=1}^{N} x_j a_{jk} x_k \right) \\
&= \int_{-\infty}^{+\infty} \left(\prod_{j=1}^{N} dx'_j \right) \exp\left(\frac{i}{\hbar} \sum_{j=1}^{N} \lambda^{(j)} x'^{\,2}_j \right) \\
&= \prod_{j=1}^{N} \frac{(i\pi\hbar)^{\frac{1}{2}}}{\sqrt{\lambda^{(j)}}} = (i\pi\hbar)^{\frac{N}{2}} \,(\det A)^{-\frac{1}{2}}\,,
\end{aligned} \tag{A.23}$$

which completes the purported proof. Notice that in the eigenvectors' basis the entanglement of the integration variables disappears, allowing for the utilization of Eq.(A.8).

By working along similar lines, it is possible to show that

$$\int_{-\infty}^{+\infty}\left(\prod_{j=1}^{N} dx_j\right)\exp\left(\frac{i}{\hbar}\sum_{j,k=1}^{N} x_j a_{jk} x_k + \frac{i}{\hbar}\sum_{j=1}^{N} b_j x_j\right)$$

$$= (i\pi\hbar)^{\frac{N}{2}}\,(\det A)^{-\frac{1}{2}}\,\exp\left[-\frac{i}{4\hbar}\sum_{j,k^1}^{N} b_j g_{jk} b_k\right], \tag{A.24}$$

where

$$g_{jk} \equiv \langle j|A^{-1}|k\rangle \tag{A.25}$$

and $b_j, j = 1,\ldots,N$ are the components of the vector $|B\rangle$ in the basis $\{|j\rangle\}$. To see how this come about we begin by recalling that

$$\sum_{j=1}^{N} b_j x_j = \sum_{j=1}^{N}\langle B|j\rangle\langle j|X\rangle = \langle B|X\rangle$$

$$= \sum_{j=1}^{N}\langle B|\lambda^{(j)}\rangle\langle\lambda^{(j)}|X\rangle = \sum_{j=1}^{N} b'_j x'_j\,. \tag{A.26}$$

Then, in view of Eqs.(A.23) and (A.10) we get

$$\int_{-\infty}^{+\infty}\left(\prod_{j=1}^{N} dx_j\right)\exp\left(\frac{i}{\hbar}\sum_{j,k=1}^{N} x_j a_{jk} x_k + \frac{i}{\hbar}\sum_{j=1}^{N} b_j x_j\right)$$

$$= \int_{-\infty}^{+\infty}\left(\prod_{j=1}^{N} dx'_j\right)\exp\left[\sum_{j=1}^{N}\left(\frac{i}{\hbar}\lambda^{(j)} x'^2_j + \frac{i}{\hbar} b'_j x'_j\right)\right]$$

$$= \prod_{j=1}^{N}\left[\frac{(i\pi\hbar)^{\frac{1}{2}}}{\sqrt{\lambda^{(j)}}}\exp\left(-\frac{i}{4\hbar}\frac{b'^2_j}{\lambda^{(j)}}\right)\right]$$

$$= (i\pi\hbar)^{\frac{N}{2}}\,(\det A)^{-\frac{1}{2}}\,\exp\left(-\frac{i}{4\hbar}\sum_{j=1}^{N}\frac{b'^2_j}{\lambda^{(j)}}\right). \tag{A.27}$$

What is left is to write this result in terms of the original variables,

$$\sum_{j=1}^{N}\frac{b'^2_j}{\lambda^{(j)}} = \sum_{j=1}^{N}\langle B|\lambda^{(j)}\rangle\frac{1}{\lambda^{(j)}}\langle\lambda^{(j)}|B\rangle = \langle B|A^{-1}|B\rangle = \sum_{j,k=1}^{N} b_j\, g_{jk}\, b_k\,, \tag{A.28}$$

which when replaced into Eq.(A.27) leads us back to Eq.(A.24).

Another integral of interest is

$$\int_{-\infty}^{+\infty}\left(\prod_{j=1}^{N} dx_j\right) F(x_1,\ldots,x_N)\exp\left(\frac{i}{\hbar}\sum_{j,k=1}^{N} x_j a_{jk} x_k\right) \tag{A.29}$$

where $F(x_1,\ldots,x_N)$ is analytic in the vicinity of $x=0$. It can, therefore, be represented by its McLauring expansion

$$F(x_1,\ldots,x_N) = \sum_{r=0}^{\infty}\frac{1}{r!}\sum_{j_1,\ldots,j_r=1}^{N}\left.\frac{\partial^r F}{\partial x_{j_1}\ldots\partial x_{j_r}}\right|_{x=0} x_{j_1}\ldots x_{j_r}\,. \tag{A.30}$$

By substituting Eq.(A.30) into Eq.(A.29) we find

$$\begin{aligned}
&\int_{-\infty}^{+\infty}\left(\prod_{j=1}^{N} dx_j\right) F(x_1,\ldots,x_N)\exp\left(\frac{i}{\hbar}\sum_{j,k=1}^{N} x_j a_{jk} x_k\right)\\
&=\sum_{r=0}^{\infty}\frac{1}{r!}\sum_{j_1,\ldots,j_r=1}^{N}\left.\frac{\partial^r F}{\partial x_{j_1}\ldots\partial x_{j_r}}\right|_{x=0}\\
&\times\int_{-\infty}^{+\infty}\left(\prod_{j=1}^{N} dx_j\right) x_{j_1}\ldots x_{j_r}\exp\left(\frac{i}{\hbar}\sum_{j,k=1}^{N} x_j a_{jk} x_k\right)\,.
\end{aligned} \tag{A.31}$$

The factors x_j entering the polynomials above will be written as

$$x_j = \frac{\hbar}{i}\frac{\partial}{\partial b_j}\exp\left(\frac{i}{\hbar}\sum_{k=1}^{N} b_k x_k\right)\Bigg|_{b=0}\,. \tag{A.32}$$

The integral in Eq.(A.29), then, yields

$$\begin{aligned}
&\int_{-\infty}^{+\infty}\left(\prod_{j=1}^{N} dx_j\right) F(x_1,\ldots,x_N)\exp\left(\frac{i}{\hbar}\sum_{j,k=1}^{N} x_j a_{jk} x_k\right)\\
&=(i\pi\hbar)^{\frac{N}{2}}(\det A)^{-\frac{1}{2}}\sum_{r=0}^{\infty}\frac{1}{r!}\sum_{j_1,\ldots,j_r=1}^{N}\left.\frac{\partial^r F}{\partial x_{j_1}\ldots\partial x_{j_r}}\right|_{x=0}\\
&\times\left(\frac{\hbar}{i}\right)^r\frac{\partial^r}{\partial b_{j_1}\ldots\partial b_{j_r}}\exp\left[-\frac{i}{4\hbar}\sum_{j,k=1}^{N} b_j g_{jk} b_k\right]\Bigg|_{b=0}\,,
\end{aligned} \tag{A.33}$$

where Eq.(A.24) was invoked.

We end this appendix by evaluating

$$\int_{-\infty}^{+\infty}\left(\prod_{j=1}^{N} dx_j\right) F\left(\frac{i}{\hbar}\sum_{k=1}^{N} t_k x_k\right)\exp\left(\frac{i}{\hbar}\sum_{j,k=1}^{N} x_j a_{jk} x_k\right). \quad \text{(A.34)}$$

Once again we switch to the eigenvectors' basis of the operator A to find

$$\int_{-\infty}^{+\infty}\left(\prod_{j=1}^{N} dx_j\right) F\left(\frac{i}{\hbar}\sum_{k=1}^{N} t_k x_k\right)\exp\left(\frac{i}{\hbar}\sum_{j,k=1}^{N} x_j a_{jk} x_k\right)$$
$$= \int_{-\infty}^{+\infty}\left(\prod_{j=1}^{N} dx'_j\right) F\left(\frac{i}{\hbar}\sum_{k=1}^{N} t'_k x'_k\right)\exp\left(\frac{i}{\hbar}\sum_{j=1}^{N} x'^2_j \lambda^{(j)}\right), \quad \text{(A.35)}$$

where

$$t'_k \equiv \langle T|\lambda^{(k)}\rangle \quad \text{(A.36)}$$

and

$$\sum_{k=1}^{N} t_k x_k = \sum_{k=1}^{N}\langle T|k\rangle\langle k|X\rangle = \langle T|X\rangle$$
$$= \sum_{k=1}^{N}\langle T|\lambda^{(k)}\rangle\langle\lambda^{(k)}|X\rangle = \sum_{k=1}^{N} t'_k x'_k . \quad \text{(A.37)}$$

We now introduce the vector $|Y\rangle$ via the dilation transformation [9]

$$x'_j = \langle\lambda^{(j)}|X\rangle \longrightarrow y_j \equiv \sqrt{\lambda^{(j)}}\, x'_j = \langle\lambda^{(j)}|Y\rangle . \quad \text{(A.38)}$$

Furthermore,

$$\prod_{j=1}^{N} dx'_j = \left(\prod_{j=1}^{N}\frac{1}{\sqrt{\lambda^{(j)}}}\right)\prod_{j=1}^{N} dy_j = (\det A)^{-\frac{1}{2}}\prod_{j=1}^{N} dy_j . \quad \text{(A.39)}$$

[9]Notice that the components of the vector $|X\rangle$ in the basis $\{|\lambda^{(k)}\rangle\}$ are not uniformly dilated.

We can then cast Eq.(A.35) as follows

$$\int_{-\infty}^{+\infty}\left(\prod_{j=1}^{N}dx_j\right)F\left(\frac{i}{\hbar}\sum_{k=1}^{N}t_kx_k\right)\exp\left(\frac{i}{\hbar}\sum_{j,k=1}^{N}x_ja_{jk}x_k\right)$$
$$= (\det A)^{-\frac{1}{2}}\int_{-\infty}^{+\infty}\left(\prod_{j=1}^{N}dy_j\right)F\left(\frac{i}{\hbar}\sum_{k=1}^{N}\frac{t'_k}{\sqrt{\lambda^{(k)}}}y_k\right)$$
$$\times\exp\left(\frac{i}{\hbar}\sum_{j=1}^{N}y_j^2\right). \tag{A.40}$$

The structure of the argument of the function F,

$$\sum_{k=1}^{N}\frac{t'_k}{\sqrt{\lambda^{(k)}}}y_k = \langle T|A^{-\frac{1}{2}}|Y\rangle\,, \tag{A.41}$$

suggests the convenience of changing to a new basis,

$$\{|\lambda^{(k)}\rangle\}\longrightarrow\{|\alpha^{(k)}\rangle\}\,, \tag{A.42}$$

which is chosen so that one of its elements, e.g., the vector $|\alpha^{(1)}\rangle$, becomes aligned with the vector $A^{-\frac{1}{2}}|T\rangle$. In that case

$$\langle T|A^{-\frac{1}{2}}|\alpha^{(k)}\rangle = c\,\delta_{1k}\,. \tag{A.43}$$

We will denote by

$$y'_k = \langle\alpha^{(k)}|Y\rangle\,,\qquad k=1,\ldots,N \tag{A.44}$$

the components of the vector $|Y\rangle$ in the basis $\{|\alpha^{(k)}\rangle\}$. Therefore, we have that

$$\langle T|A^{-\frac{1}{2}}|Y\rangle = \sum_{j=1}^{N}\langle T|A^{-\frac{1}{2}}|\alpha^{(j)}\rangle\langle\alpha^{(j)}|Y\rangle = c\,y'_1 \tag{A.45}$$

together with

$$\sum_{j=1}^{N}y_j^2 = \langle Y|Y\rangle = \sum_{j=1}^{N}{y'_j}^2\,. \tag{A.46}$$

On the other hand, the bases $\{|\lambda^{(k)}\rangle\}$ and $\{|\alpha^{(k)}\rangle\}$ can be obtained from each other through a canonical transformation. Thus,

$$\prod_{j=1}^{N} dy_j = \prod_{j=1}^{N} dy'_j \,. \tag{A.47}$$

By substituting Eqs.(A.45), (A.46) and (A.47) back into Eq.(A.40), while using Eq.(A.8) to perform the $(N-1)$ integrals not involving the variable y'_1, we obtain

$$\begin{aligned}
&\int_{-\infty}^{+\infty} \left(\prod_{j=1}^{N} dx_j\right) F\left(\frac{i}{\hbar}\sum_{k=1}^{N} t_k x_k\right) \exp\left(\frac{i}{\hbar}\sum_{j,k=1}^{N} x_j a_{jk} x_k\right) \\
&= (i\pi\hbar)^{\frac{N-1}{2}} (\det A)^{-\frac{1}{2}} \int_{-\infty}^{+\infty} du\, F\left(\frac{i}{\hbar}cu\right) e^{\frac{i}{\hbar}u^2} \\
&= (i\pi\hbar)^{\frac{N-1}{2}} \left(c^2 \det A\right)^{-\frac{1}{2}} \int_{-\infty}^{+\infty} dv\, F\left(\frac{i}{\hbar}v\right) e^{\frac{i}{\hbar}\frac{v^2}{c^2}} \,,
\end{aligned} \tag{A.48}$$

where in the last line the integration variable was changed in accordance with $u \to v \equiv cu$. This change is legitimate only if c^2 is positive definite, which can be checked to be the case from Eq.(A.43). Indeed,

$$\begin{aligned}
c^2 &= \sum_{k=1}^{N} \langle T|A^{-\frac{1}{2}}|\alpha^{(k)}\rangle\langle\alpha^{(k)}|A^{-\frac{1}{2}}|T\rangle \\
&= \langle T|A^{-1}|T\rangle = \sum_{k=1}^{N} \langle T|\lambda^{(k)}\rangle \frac{1}{\lambda^{(k)}} \langle\lambda^{(k)}|T\rangle > 0
\end{aligned} \tag{A.49}$$

as a consequence of the positivity of the operator A. Notice that in terms of the variables in Eq.(A.34)

$$c^2 = \langle T|A^{-1}|T\rangle = \sum_{j,k=1}^{N} t_j\, g_{jk}\, t_k \,. \tag{A.50}$$

Appendix B

Preliminaries on functional analysis

B.1 Functional differentiation

We begin with the definition of *functional*. Let $\mathcal{E}_C$ be the linear vector space of the complex functions $f(x)$ of the real variable $x \in [a,b]$. A functional $F[f]$ is a mapping of $\mathcal{E}_C$ into the field of complex numbers $\mathbb{C}$, namely,

$$F : f \in \mathcal{E}_C \longrightarrow c \in \mathbb{C} . \tag{B.1}$$

The integral

$$F[f] = \int_a^b dx f(x) u(x) , \tag{B.2}$$

is a particular type of functional.

The functional differential $\delta F[f]$ in terms of the functional derivative $\frac{\delta F[f]}{\delta f(x)}$ is given by

$$\delta F[f] = \int dx \, \frac{\delta F[f]}{\delta f(x)} \, \delta f(x) . \tag{B.3}$$

Meanwhile, the functional derivative is defined through the limiting process

$$\frac{\delta F[f]}{\delta f(x_0)} \equiv \lim_{\substack{\epsilon \to 0 \\ \Delta f \to 0}} \frac{F[f + \Delta f] - F[f]}{\int_{x_0 - \epsilon}^{x_0 + \epsilon} dx \, \Delta f(x)} , \tag{B.4}$$

which is illustrated in Fig. B.1. The functional derivative is taken with respect to a function f *at a certain point* x_0. Thus,

$$\frac{\delta F[f]}{\delta f(x_0)} \equiv F'[f|x_0] \tag{B.5}$$

is both a functional of f and a function of x_0. The functional argument f appears to the left of the vertical bar that divides out the bracket in $F'[f|x_0]$ while the function argument, x_0, is placed to the right of that bar. During the limiting process $\epsilon \to 0$, $\Delta f \to 0$, the point x_0 must be kept *within* the region $\Delta f \neq 0$.

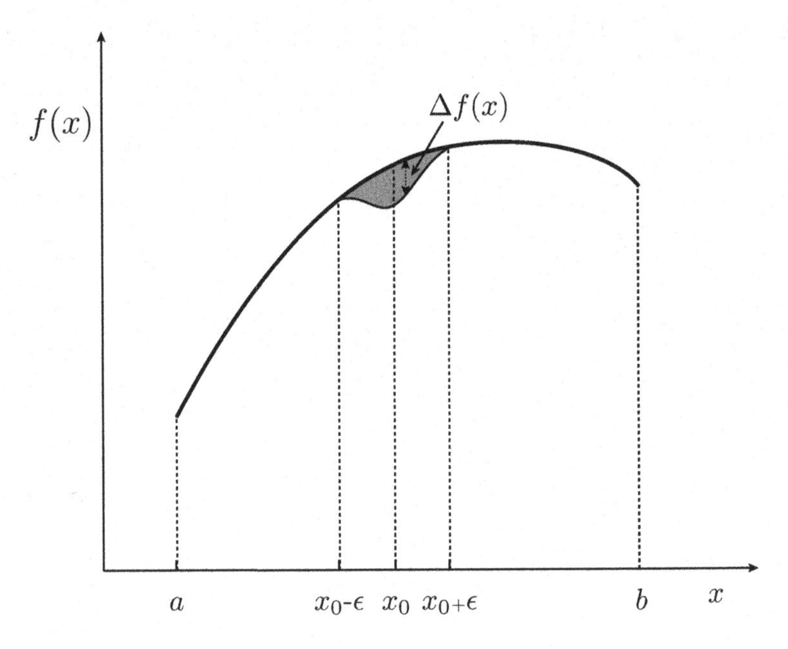

Fig. B.1 Definition of the functional derivative.

The differentiation rules for functions can be generalized to the case of functional differentiation. For instance, let

$$F = F[u]\,, \qquad u = u[v]\,. \tag{B.6}$$

Then,

$$\frac{\delta F[u]}{\delta v(x)} = \int dy\, \frac{\delta F[u]}{\delta u[v|y]}\, \frac{\delta u[v|y]}{\delta v(x)}\,, \tag{B.7}$$

exemplifies the chain rule for functional differentiation.

Consider now the problem of calculating the functional derivative

$$\frac{\delta F[f|x]}{\delta f(y_0)}$$

when

$$F[f|x] \equiv \int_{-\infty}^{+\infty} dy\, K(x,y) f(y)\,, \tag{B.8}$$

$K(x,y)$ being a continuous and symmetric kernel. Then

$$F[f + \Delta f(x)] = \int_{-\infty}^{+\infty} dy\, K(x,y)\,(f(y) + \Delta f(y)) \tag{B.9}$$

and

$$F[f + \Delta f(x)] - F[f|x] = \int_{-\infty}^{+\infty} dy\, K(x,y) \Delta f(y)\,. \tag{B.10}$$

Since $\Delta f(y)$ is nonzero only in the vicinity of y_0 we can rewrite Eq.(B.10) in the following form

$$F[f+\Delta f(x)] - F[f|x] = \int_{y_0-\epsilon}^{y_0+\epsilon} dy K(x,y) \Delta f(y) = \bar{K}(x,y_0) \int_{y_0-\epsilon}^{y_0+\epsilon} dy \Delta f(y)\,, \tag{B.11}$$

where $\bar{K}(x,y_0)$ is the mean value of $K(x,y)$ in the interval $y_0 - \epsilon \leq y \leq y_0 + \epsilon$. By substituting Eq.(B.11) into Eq.(B.4) and taking into account the continuity of $K(x,y)$ at $y = y_0$ we obtain

$$\frac{\delta F[f|x]}{\delta f(y_0)} = K(x,y_0)\,. \tag{B.12}$$

In particular, if

$$K(x,y) = \delta(x-y) \tag{B.13}$$

Eq.(B.12) yields

$$\frac{\delta f(x)}{\delta f(y)} = \delta(x-y)\,. \tag{B.14}$$

On the other hand, if $g = g(f(x))$ we have the following

$$\begin{aligned} \frac{\delta g(f(x))}{\delta f(y)} &= \frac{\partial g(f(x))}{\partial f(x)} \frac{\delta f(x)}{\delta f(y)} = \frac{\partial g(f(x))}{\partial f(x)} \delta(x-y) \\ &= \frac{\partial g(f(y))}{\partial f(y)} \delta(x-y)\,. \end{aligned} \tag{B.15}$$

Also, *ordinary and functional derivatives commute.* To verify this we consider

$$f'(x) = \int_{-\infty}^{+\infty} dy \frac{d\delta(x-y)}{dx} f(y) \,. \tag{B.16}$$

Then, by taking advantage of Eq.(B.14) we find

$$\frac{\delta f'(x)}{\delta f(z)} = \int_{-\infty}^{+\infty} dy \frac{d\delta(x-y)}{dx} \delta(y-z) = \frac{d\delta(x-z)}{dx} = \frac{d}{dx}\frac{\delta f(x)}{\delta f(z)} \,, \tag{B.17}$$

which corroborates the assertion.

B.2 Functional integrals

In this section, we present a definition for the functional integral which is not based on the *time-slicing procedure.* It applies when the functions which define the integration domain obey homogeneous boundary conditions.

B.2.1 *Functional integrals over real functions*

Consider the linear vector space $\mathcal{E}_R$ of all real square-integrable functions $\{\phi(x)|x \in [a,b]\}$ verifying the homogeneous boundary conditions

$$B_1\phi(a) = 0 \,, \tag{B.18a}$$

$$B_2\phi(b) = 0 \,. \tag{B.18b}$$

Here, B_1 and B_2 are linear differential operators. Let $\{u_j, j = 1, \ldots, \infty\}$ be an orthonormal basis in $\mathcal{E}_R$ whose components verify the boundary conditions in Eq.(B.18), i.e.,

$$B_1 u_j(a) = 0 \,, \quad \forall j \,, \tag{B.19a}$$

$$B_2 u_j(b) = 0 \,, \quad \forall j \,. \tag{B.19b}$$

We furthermore, recall that orthonormality and completeness imply, respectively, that

$$\int_a^b dx u_j(x) u_k(x) = \delta_{jk} \tag{B.20}$$

and

$$\sum_{j=1}^{\infty} u_j(x)u_j(x') = \delta(x - x') . \qquad \text{(B.21)}$$

Let $F[\phi]$ be a linear functional. The object of interest is the functional integral

$$I \equiv \int [\mathcal{D}\phi] F[\phi] , \qquad \text{(B.22)}$$

whose domain of integration includes all functions in $\mathcal{E}_R$. The expansion

$$\phi(x) = \sum_{j=1}^{\infty} C_j u_j(x) \qquad \text{(B.23)}$$

converges $\forall x$ in the closed interval $x \in [a, b]$. The functional integral in Eq.(B.22) is defined as

$$I = \int [\mathcal{D}\phi] F[\phi] \equiv \lim_{n\to\infty} \int_{-\infty}^{+\infty} dC_1 \ldots \int_{-\infty}^{+\infty} dC_n F(C_1, \ldots, C_n) . \qquad \text{(B.24)}$$

Observe that the right hand side of this last expression involves an infinite product of improper integrals.

Next we shall concentrate on the evaluation of some functional integrals of interest for our purposes. Let

$$F[\phi] = \exp\left[-\frac{1}{2} \left(\frac{2\pi}{m^2} \right) \int_a^b dx \phi^2(x) \right] , \qquad \text{(B.25)}$$

where m is a real number. From Eqs.(B.23) and (B.20) we find

$$F(C_1, \ldots, C_n) = \exp\left[-\frac{1}{2} \left(\frac{2\pi}{m^2} \right) \sum_{j=1}^{n} C_j^2 \right] . \qquad \text{(B.26)}$$

Therefore,

$$\begin{aligned} I &= \int [\mathcal{D}\phi] F[\phi] \\ &= \lim_{n\to\infty} \int_{-\infty}^{+\infty} dC_1 \ldots \int_{-\infty}^{+\infty} dC_n \exp\left[-\frac{1}{2} \left(\frac{2\pi}{m^2} \right) \sum_{j=1}^{n} C_j^2 \right] . \end{aligned} \qquad \text{(B.27)}$$

By changing integration variables

$$C_j \to A_j = \frac{\sqrt{2\pi}}{m} C_j \,, \Longrightarrow dC_j = \frac{m}{\sqrt{2\pi}} dA_j \tag{B.28}$$

Eq.(B.27) goes into

$$\begin{aligned} I &= \lim_{n\to\infty} \left[m^n \int_{-\infty}^{+\infty} \frac{dA_1}{\sqrt{2\pi}} \cdots \int_{-\infty}^{+\infty} \frac{dA_n}{\sqrt{2\pi}} \exp\left(-\frac{1}{2} \sum_{j=1}^{n} A_j^2 \right) \right] \\ &= \lim_{n\to\infty} m^n \,, \end{aligned} \tag{B.29}$$

since, according to Eq.(A.1),

$$\int_{-\infty}^{+\infty} \frac{dA}{\sqrt{2\pi}} e^{-\frac{1}{2}\alpha A^2} = \sqrt{\frac{1}{\alpha}} \,, \quad \text{for} \quad \Re\alpha > 0 \,. \tag{B.30}$$

To summarize,

$$I = \lim_{n\to\infty} m^n = \begin{cases} \infty \text{ if } m > 1 \,, \\ 1 \text{ if } m = 1 \,, \\ 0 \text{ if } m < 1 \,, \end{cases} \tag{B.31}$$

so that

$$I_1 \equiv \int \left[\frac{\mathcal{D}\phi}{\sqrt{2\pi}} \right] e^{-\frac{1}{2}\int_a^b dx \phi^2(x)} = 1 \,, \tag{B.32}$$

where the functional measure is defined as

$$\left[\frac{\mathcal{D}\phi}{\sqrt{2\pi}} \right] \equiv \lim_{n\to\infty} \prod_{j=1}^{n} \frac{dC_j}{\sqrt{2\pi}} \,. \tag{B.33}$$

The result found in Eq.(B.32) can be generalized to

$$I_2 \equiv \int \left[\frac{\mathcal{D}\phi}{\sqrt{2\pi m}} \right] e^{-\frac{1}{2}\left(\frac{1}{m}\right)\int_a^b dx \phi^2(x)} = 1 \,, \tag{B.34}$$

where

$$\left[\frac{\mathcal{D}\phi}{\sqrt{2\pi m}} \right] \equiv \lim_{n\to\infty} \prod_{j=1}^{n} \frac{dC_j}{\sqrt{2\pi m}} \,. \tag{B.35}$$

Equation (A.3) can be used to justify that

$$\int \left[\frac{\mathcal{D}\phi}{\sqrt{2\pi}}\right] e^{-\frac{1}{2}\int_a^b dx\phi^2(x)} = \int \left[\frac{\mathcal{D}\phi}{\sqrt{2\pi}}\right] e^{-\frac{1}{2}\int_a^b dx[\phi(x)+\eta(x)]^2} = 1\,, \tag{B.36}$$

where $\eta(x) \in \mathcal{E}_R$. From now on we shall assume that

i) functional integrals are *translationally invariant*, i.e.,

$$\int [\mathcal{D}\phi] F[\phi] = \int [\mathcal{D}\phi] F[\phi + \eta]\,. \tag{B.37}$$

ii) functional integrals *depend linearly on their integrands*. Therefore, the differentiation under the sign of functional integration can be performed in accordance with the usual rules of calculus.

Below, we focus on the evaluation of the functional integral

$$I_3 \equiv \int \left[\frac{\mathcal{D}\phi}{\sqrt{2\pi}}\right] e^{-\frac{1}{2}\int_a^b dx\phi^2(x) - \int_a^b dx\phi(x)\eta(x)}\,. \tag{B.38}$$

For this purpose, we shall begin by writing

$$-\frac{1}{2}\phi^2(x) - \phi(x)\eta(x) = -\frac{1}{2}\left[\phi(x) + \eta(x)\right]^2 + \frac{1}{2}\eta^2(x)\,, \tag{B.39}$$

which if put back into Eq.(B.38) enables us to find

$$I_3 = e^{\frac{1}{2}\int_a^b dx\eta^2(x)} \int \left[\frac{\mathcal{D}\phi}{\sqrt{2\pi}}\right] e^{-\frac{1}{2}\int_a^b dx[\phi(x)+\eta(x)]^2} = e^{\frac{1}{2}\int_a^b dx\eta^2(x)}\,, \tag{B.40}$$

where Eqs.(B.37) and (B.32) were used. We highlight that this result is a consequence of the translational invariance of the functional integral.

Coming next is the functional integral

$$I_4 \equiv \int \left[\frac{\mathcal{D}\phi}{\sqrt{2\pi}}\right] F[\phi] e^{-\frac{1}{2}\int_a^b dx\phi^2(x)}\,, \tag{B.41}$$

where $F[\phi]$ is analytic at $\phi = 0$. We can, therefore, invoke the functional McLauring expansion

$$F[\phi] = \sum_{n=0}^{\infty} \frac{1}{n!} \int_a^b dx_1 \cdots \int_a^b dx_n \left.\frac{\delta^n F[\phi]}{\delta\phi(x_1)\cdots\delta\phi(x_n)}\right|_{\phi=0} \phi(x_1)\cdots\phi(x_n)\,, \tag{B.42}$$

which allows us to write

$$I_4 = \sum_{n=0}^{\infty} \frac{1}{n!} \int_a^b dx_1 \cdots \int_a^b dx_n \left. \frac{\delta^n F[\phi]}{\delta\phi(x_1)\cdots\delta\phi(x_n)} \right|_{\phi=0}$$
$$\times \int \left[\frac{\mathcal{D}\phi}{\sqrt{2\pi}}\right] \phi(x_1)\cdots\phi(x_n) e^{-\frac{1}{2}\int_a^b dx \phi^2(x)} . \quad \text{(B.43)}$$

On the other hand,

$$\phi(x) = -\frac{\delta}{\delta J(x)} \exp\left[-\int_a^b dx \phi(x) J(x)\right]\Bigg|_{J=0} . \quad \text{(B.44)}$$

Hence, Eq.(B.43) can be cast

$$I_4 = \sum_{n=0}^{\infty} \frac{1}{n!} \int_a^b dx_1 \cdots \int_a^b dx_n \left. \frac{\delta^n F[\phi]}{\delta\phi(x_1)\cdots\delta\phi(x_n)} \right|_{\phi=0}$$
$$\times (-1)^n \frac{\delta^n}{\delta J(x_1)\cdots\delta J(x_n)} \int \left[\frac{\mathcal{D}\phi}{\sqrt{2\pi}}\right] e^{-\frac{1}{2}\int_a^b dx \phi^2(x) - \int_a^b dx \phi(x) J(x)} \Bigg|_{J=0}$$
$$= \sum_{n=0}^{\infty} \frac{1}{n!} \int_a^b dx_1 \cdots \int_a^b dx_n \left. \frac{\delta^n F[\phi]}{\delta\phi(x_1)\cdots\delta\phi(x_n)} \right|_{\phi=0}$$
$$\times (-1)^n \frac{\delta^n}{\delta J(x_1)\cdots\delta J(x_n)} e^{\frac{1}{2}\int_a^b dx J^2(x)} \Bigg|_{J=0}$$
$$= F\left[-\frac{\delta}{\delta J}\right] e^{\frac{1}{2}\int_a^b dx J^2(x)} \Bigg|_{J=0} . \quad \text{(B.45)}$$

This book refers to the function $J(x)$ as the *fictitious source* for the variable $\phi(x)$. It belongs to $\mathcal{E}_R$.

So far, we have been restricted to study Gaussian functional integrals. Their integrand includes a damping factor that causes the limit operation in Eq.(B.24) to smooth. On the other hand, the functional formulation of quantum mechanics demands the evaluation of functional integrals that exhibit an oscillatory factor instead. For instance, we will be required to compute

$$I_5 \equiv \int \left[\frac{\mathcal{D}\phi}{\sqrt{2\pi i}}\right] e^{\frac{i}{2}\int_a^b dx \phi^2(x)} . \quad \text{(B.46)}$$

By applying the definition in Eq.(B.24) we find

$$I_5 = \lim_{n\to\infty} \left[\int_{-\infty}^{+\infty} \frac{dA_1}{\sqrt{2\pi i}} \cdots \int_{-\infty}^{+\infty} \frac{dA_n}{\sqrt{2\pi i}} \exp\left(\frac{i}{2}\sum_{j=1}^{n} A_j^2\right)\right] . \quad \text{(B.47)}$$

Each integral in the right hand side of Eq.(B.47) will be evaluated according to the prescription we introduced and discussed in Appendix A. This systematics leads to

$$I_5 = \int \left[\frac{\mathcal{D}\phi}{\sqrt{2\pi i}}\right] e^{\frac{i}{2}\int_a^b dx \phi^2(x)} = 1\,. \tag{B.48}$$

The translational invariance of the functional integral, as well as the linear dependence upon its integrand, are assumed to remain valid. We can show that the counterparts of Eqs.(B.38) and (B.41) are, respectively,

$$I_6 \equiv \int \left[\frac{\mathcal{D}\phi}{\sqrt{2\pi}}\right] e^{\frac{i}{2}\int_a^b dx \phi^2(x) + i\int_a^b dx \phi(x)\eta(x)} = e^{-\frac{i}{2}\int_a^b dx \eta^2(x)} \tag{B.49}$$

and

$$I_7 \equiv \int \left[\frac{\mathcal{D}\phi}{\sqrt{2\pi}}\right] F[\phi] e^{\frac{i}{2}\int_a^b dx \phi^2(x)} = F\left[\frac{1}{i}\frac{\delta}{\delta J}\right] e^{-\frac{i}{2}\int_a^b dx J^2(x)}\Bigg|_{J=0} . \tag{B.50}$$

B.2.2 *Functional integrals over complex functions*

Let $\alpha(x)$ and $\beta(x)$ be functions in $\mathcal{E}_R$. From the results in subsection B.2.1 it follows that

$$\int \left[\frac{\mathcal{D}\alpha}{\sqrt{2\pi}}\right]\left[\frac{\mathcal{D}\beta}{\sqrt{2\pi}}\right] e^{-\frac{1}{2}\int_a^b dx \alpha^2(x) - \frac{1}{2}\int_a^b dx \beta^2(x)} = 1\,. \tag{B.51}$$

We define the *complex* function ϕ of the *real* variable x as

$$\phi(x) = \frac{1}{\sqrt{2}}(\alpha + i\beta)\,, \quad \phi^\star(x) = \frac{1}{\sqrt{2}}(\alpha - i\beta)\,. \tag{B.52}$$

We also introduce the complex functional integration measure by means of

$$\left[\frac{\mathcal{D}\phi}{\sqrt{2\pi}}\right]\left[\frac{\mathcal{D}\phi^\star}{\sqrt{2\pi}}\right] \equiv \left[\frac{\mathcal{D}\alpha}{\sqrt{2\pi}}\right]\left[\frac{\mathcal{D}\beta}{\sqrt{2\pi}}\right]. \tag{B.53}$$

With these ingredients at hand, it becomes possible to introduce the functional integral over complex functions by rewriting (B.51) in the following form

$$I_8 = \int \left[\frac{\mathcal{D}\phi}{\sqrt{2\pi}}\right]\left[\frac{\mathcal{D}\phi^\star}{\sqrt{2\pi}}\right] e^{-\int_a^b dx \phi^\star(x)\phi(x)} = 1\,. \tag{B.54}$$

In the case of an oscillatory integrand, the counterpart is

$$I_9 = \int \left[\frac{\mathcal{D}\phi}{\sqrt{2\pi i}}\right] \left[\frac{\mathcal{D}\phi^\star}{\sqrt{2\pi i}}\right] e^{i\int_a^b dx \phi^\star(x)\phi(x)} = 1\,. \tag{B.55}$$

In order to formulate the perturbative expansion regarding complex functions we first need to evaluate

$$I_{10} = \int \left[\frac{\mathcal{D}\phi}{\sqrt{2\pi i}}\right] \left[\frac{\mathcal{D}\phi^\star}{\sqrt{2\pi i}}\right] e^{i\int_a^b dx \phi^\star(x)\phi(x) + i\int_a^b dx \phi^\star(x)J(x) + i\int_a^b dx J^\star(x)\phi(x)}\,, \tag{B.56}$$

where J and $J^\star$ are the fictitious sources for $\phi^\star$ and ϕ, respectively. To this end, we first substitute

$$\phi^\star\phi + J^\star\phi + \phi^\star J = (\phi^\star + J^\star)(\phi + J) - J^\star J \tag{B.57}$$

into Eq.(B.56) and then invoke the translational invariance of the functional integral in order to obtain

$$\begin{aligned} I_{10} &= \int \left[\frac{\mathcal{D}\phi}{\sqrt{2\pi i}}\right] \left[\frac{\mathcal{D}\phi^\star}{\sqrt{2\pi i}}\right] e^{i\int_a^b dx \phi^\star(x)\phi(x) + i\int_a^b dx \phi^\star(x)J(x) + i\int_a^b dx J^\star(x)\phi(x)} \\ &= e^{-i\int_a^b dx J^\star(x)J^\star(x)}\,. \end{aligned} \tag{B.58}$$

It is straightforward to verify that

$$\begin{aligned} I_{11} &= \int \left[\frac{\mathcal{D}\phi}{\sqrt{2\pi i}}\right] \left[\frac{\mathcal{D}\phi^\star}{\sqrt{2\pi i}}\right] F[\phi^\star, \phi]\, e^{i\int_a^b dx \phi^\star(x)\phi(x)} \\ &= F\left[\frac{1}{i}\frac{\delta}{\delta J}, \frac{1}{i}\frac{\delta}{\delta J^\star}\right] e^{-i\int_a^b dx J^\star(x)J^\star(x)}\Bigg|_{J=J^\star=0}\,, \end{aligned} \tag{B.59}$$

where $F[\phi^\star, \phi]$ is a functional of ϕ and $\phi^\star$ analytic at $\phi = \phi^\star = 0$.

B.2.3 *Change of variables in the functional integral*

We now go back to the functional integrals over real variables. We hence face the task of evaluating the functional integral

$$I_{12} = \int \left[\frac{\mathcal{D}\phi}{\sqrt{2\pi i}}\right] e^{\frac{i}{2}\int_a^b dx \int_a^b dy\, \phi(x)K(x,y)\phi(y)}\,, \tag{B.60}$$

where

$$K(x, y) = K(y, x) \tag{B.61}$$

is a real and symmetric kernel. The domain of integration remains as before. The kernel $K(x, y)$ possesses, by assumption, a complete set of orthonormal eigenfunctions $\{u_j(x)\}$,

$$\int_a^b dx\, K(x,y) u_j(y) = \lambda_j u_j(x)\,, \tag{B.62}$$

serving as basis in $\mathcal{E}_R$. The Hermitian nature of $K(x, y)$ insures that the eigenvalues $\{\lambda_j\}$ are real. We shall also assume that they are all *positive definite.* By using the eigenfunctions of $K(x, y)$ to expand the functions $\{\phi(x)\}$ we find

$$\int_a^b dx \int_a^b dy \phi(x) K(x,y) \phi(y) = \sum_{j=1}^{\infty} \lambda_j C_j^2\,, \tag{B.63}$$

where Eq.(B.20) was used. Thus, the definition of functional integral adopted here leads us to

$$I_{12} = \lim_{n\to\infty} \int_{-\infty}^{+\infty} \left(\prod_{j=1}^{n} \frac{dC_j}{\sqrt{2\pi i}} \right) \exp\left(\frac{i}{2} \sum_{j=1}^{n} \lambda_j C_j^2 \right) . \tag{B.64}$$

The positivity of the eigenvalues λ_j validates the change of variables

$$\sqrt{\lambda_j} C_j = B_j\,, \Longrightarrow dC_j = \frac{dB_j}{\sqrt{\lambda_j}}\,, \tag{B.65a}$$

$$C_j \longrightarrow -\infty \Longrightarrow B_j \longrightarrow -\infty\,, \tag{B.65b}$$

$$C_j \longrightarrow +\infty \Longrightarrow B_j \longrightarrow +\infty\,. \tag{B.65c}$$

Hence,

$$\begin{aligned} I_{12} &= \int \left[\frac{\mathcal{D}\phi}{\sqrt{2\pi i}} \right] e^{\frac{i}{2} \int_a^b dx \int_a^b dy\, \phi(x) K(x,y) \phi(y)} \\ &= \left(\lim_{n\to\infty} \prod_{j=1}^{n} \frac{1}{\sqrt{\lambda_j}} \right) I_5 = (\det K)^{-\frac{1}{2}}\,, \end{aligned} \tag{B.66}$$

where the determinant of the operator K is given by

$$\det K \equiv \prod_{j=1}^{\infty} \lambda_J\,. \tag{B.67}$$

A closely related integral is

$$I_{13} = \int \left[\frac{\mathcal{D}\phi}{\sqrt{2\pi i}} \right] e^{\frac{i}{2} \int_a^b dx \int_a^b dy \phi(x) K(x,y) \phi(y) + i \int_a^b dx \phi(x) J(x)} . \tag{B.68}$$

We then invoke $J(x) \in \mathcal{E}_R$ to write

$$J(x) = \sum_{j=1}^{\infty} D_j u_j(x) , \tag{B.69}$$

whose coefficients are calculated from

$$D_j = \int_a^b dx u_j(x) J(x) . \tag{B.70}$$

Therefore, the definition of the functional integral in Eq.(B.24) leads to

$$I_{13} = \lim_{n \to \infty} \int_{-\infty}^{+\infty} \left(\prod_{j=1}^{n} \frac{dC_j}{\sqrt{2\pi i}} \right) \exp \left(\frac{i}{2} \sum_{j=1}^{n} \lambda_j C_j^2 + i \sum_{j=1}^{n} C_j D_j \right) . \tag{B.71}$$

The change of integration variables defined in Eq.(B.65) yields

$$\begin{aligned} I_{13} = & \left(\lim_{n \to \infty} \prod_{j=1}^{n} \frac{1}{\sqrt{\lambda_j}} \right) \lim_{n \to \infty} \int_{-\infty}^{+\infty} \left(\prod_{j=1}^{n} \frac{dB_j}{\sqrt{2\pi i}} \right) \\ & \times \exp \left(\frac{i}{2} \sum_{j=1}^{n} B_j^2 + i \sum_{j=1}^{n} B_j \frac{1}{\sqrt{\lambda_j}} D_j \right) . \end{aligned} \tag{B.72}$$

However,

$$\frac{1}{2} B_j^2 + B_j \frac{1}{\sqrt{\lambda_j}} D_j = \frac{1}{2} \left(B_j + \frac{1}{\sqrt{\lambda_j}} D_j \right)^2 - \frac{1}{2} D_j \frac{1}{\lambda_j} D_j . \tag{B.73}$$

By then appealing to translational invariance we obtain

$$I_{13} = \left(\lim_{n \to \infty} \prod_{j=1}^{n} \frac{1}{\sqrt{\lambda_j}} \right) \left[\lim_{n \to \infty} \exp \left(-\frac{i}{2} \sum_{j=1}^{n} D_j \frac{1}{\lambda_j} D_j \right) \right] . \tag{B.74}$$

The fact that all the eigenvalues of the operator $K(x,y)$ are positive definite secures the existence and uniqueness of its inverse (Δ_K),

$$\int_a^b dy K(x,y)\Delta_K(y,z) = \delta(x-z)\,. \tag{B.75}$$

The spectral resolution of Δ_K in terms of the eigenfunctions and eigenvalues of $K(x,y)$ reads

$$\Delta_K(x,y) = \sum_{j=1}^{\infty} u_j(x)\frac{1}{\lambda_j}u_j(y)\,. \tag{B.76}$$

On the other hand, by using equation Eq.(B.70) we find

$$\begin{aligned}\sum_{j=1}^{\infty} D_j \frac{1}{\lambda_j} D_j &= \int_a^b dx \int_a^b dy J(x) \left[\sum_{j=1}^{\infty} u_j(x)\frac{1}{\lambda_j}u_j(y)\right] J(y)\\ &= \int_a^b dx \int_a^b dy J(x)\Delta_K(x,y)J(y)\,. \end{aligned} \tag{B.77}$$

Then, by returning with Eq.(B.77) into Eq.(B.74) we arrive at

$$\begin{aligned} I_{13} &= \int \left[\frac{\mathcal{D}\phi}{\sqrt{2\pi i}}\right] e^{\frac{i}{2}\int_a^b dx \int_a^b dy \phi(x)K(x,y)\phi(y) + i\int_a^b dx \phi(x)J(x)}\\ &= (\det K)^{-\frac{1}{2}} \exp\left[-\frac{i}{2}\int_a^b dx \int_a^b dy J(x)\Delta_K(x,y)J(y)\right]\,. \end{aligned} \tag{B.78}$$

As a by-product of this result we observe that

$$\begin{aligned} I_{14} &= \int \left[\frac{\mathcal{D}\phi}{\sqrt{2\pi i}}\right] F[\phi] e^{\frac{i}{2}\int_a^b dx \int_a^b dy \phi(x)K(x,y)\phi(y)} = (\det K)^{-\frac{1}{2}}\\ &\times F\left[\frac{1}{i}\frac{\delta}{\delta J}\right] \exp\left[-\frac{i}{2}\int_a^b dx \int_a^b dy J(x)\Delta_K(x,y)J(y)\right]\Bigg|_{J=0} \end{aligned} \tag{B.79}$$

and, in particular,

$$\begin{aligned} I_{15} &= \int \left[\frac{\mathcal{D}\phi}{\sqrt{2\pi i}}\right] \phi(x)\phi(y) e^{\frac{i}{2}\int_a^b dz \int_a^b dz' \phi(z)K(z,z')\phi(z')}\\ &= (\det K)^{-\frac{1}{2}} \frac{1}{i}\frac{\delta}{\delta J(x)}\frac{1}{i}\frac{\delta}{\delta J(y)} \exp\left[-\frac{i}{2}\int_a^b dz \int_a^b dz' J(z)\Delta_K(z,z')J(z')\right]\Bigg|_{J=0}\\ &= (\det K)^{-\frac{1}{2}}\, i\Delta_K(x,y)\,. \end{aligned} \tag{B.80}$$

For functional integrals defined over complex functions, the primary object of interest is

$$I_{16} = \int \left[\frac{\mathcal{D}\phi}{\sqrt{2\pi i}}\right] \left[\frac{\mathcal{D}\phi^\star}{\sqrt{2\pi i}}\right] e^{i \int_a^b dx \int_a^b dy \phi^\star(x) K(x,y) \phi(y)} , \tag{B.81}$$

where the assumptions made in connection with $K(x, y)$ still apply. By calling upon Eq.(B.52) we find

$$\begin{aligned} &\int_a^b dx \int_a^b dy \phi^\star(x) K(x,y) \phi(y) \\ &= \frac{1}{2} \int_a^b dx \int_a^b dy \alpha(x) K(x,y) \alpha(y) \\ &+ \frac{1}{2} \int_a^b dx \int_a^b dy \beta(x) K(x,y) \beta(y) , \end{aligned} \tag{B.82}$$

since

$$\frac{i}{2} \int_a^b dx \int_a^b dy \left[\alpha(x) K(x,y) \beta(y) - \beta(x) K(x,y) \alpha(y)\right] = 0 . \tag{B.83}$$

Then, Eq.(B.66) implies that

$$I_{16} = \int \left[\frac{\mathcal{D}\phi}{\sqrt{2\pi i}}\right] \left[\frac{\mathcal{D}\phi^\star}{\sqrt{2\pi i}}\right] e^{i \int_a^b dx \int_a^b dy \phi^\star(x) K(x,y) \phi(y)} = (\det K)^{-1} . \tag{B.84}$$

It is straightforward to verify that

$$\begin{aligned} I_{17} &= \int \left[\frac{\mathcal{D}\phi}{\sqrt{2\pi i}}\right] \left[\frac{\mathcal{D}\phi^\star}{\sqrt{2\pi i}}\right] \\ &\times \exp\left\{ i \int_a^b dx \int_a^b dy \phi^\star(x) K(x,y) \phi(y) + i \int_a^b dx \left[J^\star(x)\phi(x) + \phi^\star(x) J(x)\right] \right\} \\ &= (\det K)^{-1} \exp\left[-\frac{i}{2} \int_a^b dx \int_a^b dy J^\star(x) \Delta_K(x,y) J(y) \right] \end{aligned} \tag{B.85}$$

and

$$\begin{aligned} I_{18} &= \int \left[\frac{\mathcal{D}\phi}{\sqrt{2\pi i}}\right] \left[\frac{\mathcal{D}\phi^\star}{\sqrt{2\pi i}}\right] F\left[\phi^\star, \phi\right] \exp\left\{ i \int_a^b dx \int_a^b dy \phi^\star(x) K(x,y) \phi(y) \right\} \\ &= (\det K)^{-1} F\left[\frac{1}{i}\frac{\delta}{\delta J}, \frac{1}{i}\frac{\delta}{\delta J^\star}\right] \\ &\times \exp\left[-\frac{i}{2} \int_a^b dx \int_a^b dy J^\star(x) \Delta_K(x,y) J(y) \right] \Bigg|_{J=J^\star=0} . \end{aligned} \tag{B.86}$$

B.2.4 *Functional Fourier transform*

Let $\alpha(x) \in \mathcal{E}_R$ and $F[\alpha]$ be a linear functional. We will denote by $\tilde{F}[\omega]$ the functional Fourier transform of $F[\alpha]$. By definition,

$$F[\alpha] = \int \left[\frac{\mathcal{D}\omega}{2\pi}\right] e^{i \int_a^b dx \alpha(x)\omega(x)} \tilde{F}[\omega] \tag{B.87}$$

and

$$\tilde{F}[\omega] = \int [\mathcal{D}\beta] \, e^{-i \int_a^b dx \beta(x)\omega(x)} F[\beta] \,. \tag{B.88}$$

By bringing Eq.(B.88) into Eq.(B.87) we obtain an expression for the generalized Dirac delta functional

$$\delta[\alpha - \beta] = \int \left[\frac{\mathcal{D}\omega}{2\pi}\right] e^{i \int_a^b dx [\alpha(x) - \beta(x)]\omega(x)} \,. \tag{B.89}$$

As expected, we can check that

$$F[\alpha] = \int [\mathcal{D}\beta] \, \delta[\alpha - \beta] F[\beta] \,. \tag{B.90}$$

We turn now into obtaining an explicit expression for the delta functional in terms of ordinary delta functions. To this end we start by making use of the expansions

$$\alpha(x) = \sum_{j=1}^{\infty} a_j u_j(x) \,, \tag{B.91a}$$

$$\beta(x) = \sum_{j=1}^{\infty} b_j u_j(x) \,, \tag{B.91b}$$

$$\omega(x) = \sum_{j=1}^{\infty} c_j u_j(x) \,. \tag{B.91c}$$

Accordingly, we find

$$\delta[\alpha-\beta] = \lim_{n\to\infty} \int \prod_{j=1}^{n} \left(\frac{dc_j}{2\pi}\right) e^{i \sum_{j=1}^{n} c_j (a_j - b_j)} = \lim_{n\to\infty} \prod_{j=1}^{n} \delta(a_j - b_j) \,, \tag{B.92}$$

which is the desired expression.

At last, we study the delta functional

$$\delta[\alpha]\,, \tag{B.93}$$

when

$$\alpha(x) = \int_a^b dx K(x,y)\beta(y)\,, \tag{B.94}$$

and K is the above defined operator. In view of Eq.(B.89) one has that

$$\delta[\alpha] = \int \left[\frac{\mathcal{D}\omega}{2\pi}\right] e^{i\int_a^b dx \int_a^b dy \omega(x) K(x,y) \beta(y)}\,, \tag{B.95}$$

where Eq.(B.94) has been used. By recalling the expansions in Eq.(B.91) we achieve

$$\int_a^b dx \int_a^b dy \omega(x) K(x,y) \beta(y) = \sum_{j=1}^{\infty} c_j \lambda_j a_j \tag{B.96}$$

and, as a result,

$$\delta[\alpha] = \lim_{n\to\infty} \int_{-\infty}^{+\infty} \left(\prod_{j=1}^{n} \frac{dc_j}{2\pi}\right) e^{\sum_{j=1}^{\infty} c_j \lambda_j a_j} = (\det K)^{-1}\, \delta[\beta]\,. \tag{B.97}$$

Bibliography

Abers, E. S. and Lee, B. W. (1973). Gauge theories, *Phys. Rep.* **9**, 1, pp. 1–141.
Agarwal, G. S. and Wolf, E. (1970). Calculus for functions of noncommuting operators and general phase space methods in quantum mechanics. I. Mapping theorems and ordering of functions of noncommuting operators, *Phys. Rev. D* **2**, 10, pp. 2161–2186.
Bemfica, F. S. (2009). *Dinâmica Quântica de Sistemas Não-Comutativos*, Ph.D. thesis, Universidade Federal do Rio Grande do Sul, Rio Grande do Sul - Brazil.
Bemfica, F. S. and Girotti, H. O. (2005). The noncommutative degenerate electron gas, *J. Phys. A: Math. Gen.* **38**, pp. L539–L547.
Bemfica, F. S. and Girotti, H. O. (2008a). Noncommutative quantum mechanics: Uniqueness of the functional description, *Phys. Rev. D* **78**, 12, p. 125009.
Bemfica, F. S. and Girotti, H. O. (2008b). On the quantum dynamics of noncommutative systems, *Braz. J. Phys* **38**, 2, pp. 227–236.
Biggati, D. and Susskind, L. (2000). Magnetic fields, branes and noncommutative geometry, *Phys. Rev. D* **62**, 6, p. 066004.
Chaichian, S.-J. M. M., M. and Tureanu, A. (2004). Non-commutativity of space-time and the hydrogen spectrum atom, *Eur. Phys. J. C* **36**, 2, pp. 251–252.
Christ, N. H. and Lee, T. D. (1980). Operator ordering and Feynman rules in gauge theories, *Phys. Rev. D* **22**, 4, pp. 939–958.
Cohen, L. (1966). Generalized phase space distributions, *J. Math. Phys.* **7**, 5, pp. 781–786.
Cohen, L. (1976). Hamiltonian operators via Feynman path integrals, *J. Math. Phys.* **17**, 4, pp. 597–598.
Coleman, S. and Weinberg, E. (1973). Radiative corrections as the origin of spontaneous symmetry breaking, *Phys. Rev.* **7**, 6, p. 1888.
Connes, M. R., A. Douglas and Schwarz, A. (1998). Noncommutative geometry and Matrix theory, *JHEP* **1998**, 02, p. 003.
Costa, G. H. O., M. E. V and Simões, T. J. M. (1985). Dynamics of gauge systems and Dirac conjecture, *Phys. Rev. D* **32**, 2, pp. 405–410.
Costa, M. E. V. and Girotti, H. O. (1981). Quantization of gauge-invariant theories through the Dirac bracket formalism, *Phys. Rev. D* **24**, 12, pp. 3323–

3325.

Deriglazov, A. A. (2002). Noncommutative version of an arbitrary nondegenerate mechanics, Tech. Rep. hep-th/0208072.

Dirac, P. A. M. (1933). The Lagrangian in quantum mechanics, *Physk. Zeits. Sowjetunion* **3**, 1, pp. 64–72.

Dirac, P. A. M. (1964). *Lectures in Quantum Mechanics* (Belfer Graduate School of Science, New York).

Douglas, M. R. and Nekrasov, N. A. (2001). Noncommutative field theory, *Rev. Mod. Phys.* **73**, 4, pp. 977–1029.

Edwards, S. F. and Gulyaev, Y. V. (1964). Path integrals in polar coordinates, *P. Roy. Soc. Lond. A* **279**, 1377, pp. 229–235.

Erdélyi, A., Magnus, W., Oberhettinger, F. G. and Tricomi, F. G. (1953). *Higher Trascendental Functions* (Mc Graw-Hill, Inc., New York).

Faddeev, L. D. (1970). The Feynman integral for singular Lagrangians, *Theor. Math. Phys.* **1**, 1, pp. 1–13.

Faddeev, L. D. and Popov, V. N. (1967). Feynman diagrams for the Yang-Mills field, *Phys. Lett.* **25B**, 1, p. 29.

Feynman, R. P. (1948). Space-time approach to non-relativistic quantum mechanics, *Rev. Mod. Phys.* **20**, 2, pp. 367–387.

Feynman, R. P. (1950). Mathematical formulation of the quantum theory of electromagnetic interaction, *Phys. Rev.* **80**, 3, pp. 440–457.

Feynman, R. P. and Hibbs, A. R. (1965). *Quantum Mechanics and Path Integrals* (McGraw-Hill, Inc., New York).

Filk, T. (1996). Divergencies in a field theory on quantum space, *Phys. Lett.* **376B**, 1/3, pp. 53–58.

Fradkin, E. S. and Vilkovisky, G. A. (1975). Quantization of relativistic systems with constraints, *Phys. Lett.* **55B**, 2, pp. 224–226.

Fradkin, E. S. and Vilkovisky, G. A. (1977). Quantization of Relativistic Systems with Constraints. Equivalence of Canonical and Covariant Formalisms in Quantum Theory of Gravitational Field, Theoretical Physics Ref.TH.2332-CERN, CERN, Geneva - Switzerland.

Gamboa, L. M., J. and Rojas, J. C. (2001). Noncommutative quantum mechanics, *Phys. Rev. D* **64**, 6, p. 067901.

Gamboa, M. M. F., J. Loewe and Rojas, J. C. (2002). Noncommutative quantum mechanics: the twp-dimensional central field, *Int. J. Mod. Phys. A* **17**, 19, pp. 2555–2565.

Garrod, C. (1966). Hamiltonian path-integral methods, *Rev. Mod. Phys.* **38**, 3, pp. 483–494.

Gervais, J. L. and Jevicki, A. (1976). Point canonical transformations in the path integral, *Nucl. Phys. B* **110**, 1, pp. 93–112.

Girotti, H. O. (1990). *Classical and Quantum Dynamics of Constrained Systems*, Jorge Andre Swieca Summer School (World Scientific), pp. 1–77.

Girotti, H. O. (2003). Noncommutative quantum field theories, Talk at the 12 jorge andre swieca summer school: Section particle and fields, Sociedade Brasileira de Fisica, Campos de Jordao, Brazil.

Girotti, H. O. (2004). Noncommutative quantum mechanics, *Am. J. Phys.* **72**, 5,

pp. 608–612.
Girotti, H. O. and Simões, T. J. M. (1980). Uniqueness of the functional approach, *Phys. Rev. D* **22**, 6, pp. 1385–1389.
Gitman, D. M. and Tyutin, I. V. (1990). *Quantization of Fields with Constraints* (Springer-Verlag, Berlin).
Gomes, M. (2001). Renormalization in noncommutative field theory, Talk at the 11 Jorge Andre Swieca Summer School: Section particle and fields, Sociedade Brasileira de Fisica, Campos de Jordao, Brazil.
Gradshteyn, I. S. and Ryzhik, I. M. (1980). *Table of Integrals, Series, and Products* (Academic Press, Inc., New York).
Grönewold, H. J. (1946). On the principles of elementary quantum mechanics, *Physica* **12**, 7, pp. 405–460.
Henneaux, M. and Teitelboim, C. (1992). *Quantization of Gauge Systems* (Princeton University Press, Princeton, New Jersey).
Horváthy, P. A. and Plyushchay, M. S. (2002). Non-relativistic anyons, exotic galilean symmetry and noncommutative plane, *JHEP* **2002**, 06, p. 033.
Jackiw, R. (2001). Physical instances of noncommuting coordinates, Talk at the School on String Theory hep-th/0110057, Istanbul, Turkey.
Landau, L. D. and Lifshitz, E. M. (1958). *Quantum Mechanics - Non-Relativistic Theory* (Pergamon Press, London).
Leschke, H. and Schmutz, M. (1977). Operator orderings and functional formulations of quantum and stochastical mechanics, *Z. Phys. B* **27**, 1, pp. 85–94.
Menzinescu, L. (2000). Star Operation in Quantum Mechanics, Tech. Rep. hep-th/0007046.
Messiah, A. (1966). *Quantum Mechanics* (John Wiley and Sons, New York).
Mizrahi, M. M. (1975). The Weyl correspondence and path integrals, *J. Math. Phys* **16**, 11, pp. 2201–2206.
Moyal, J. E. (1949). Quantum mechanics as a statistical theory, *Proc. Cambridge Philos. Soc* **45**, 1, pp. 99–124.
Ramond, P. (1990). *Field Theory: A Modern Primer* (Addison Wesley Publishing Company, Inc., New York).
Rivelles, V. (2000). Noncommutative supersymmetric field theories, *Braz. J. Phys* **31**, 2, pp. 255–262.
Schwinger, J. (1951a). On the Green's functions of quantized fields.I, *P. Natl. Acad. Sci.* **37**, pp. 452–455.
Schwinger, J. (1951b). On the Green's functions of quantized fields.II, *P. Natl. Acad. Sci.* **37**, pp. 455–459.
Schwinger, J. (1951c). The theory of quantized fields. I, *Phys. Rev.* **82**, 4, pp. 914–927.
Seiberg, N. and Witten, E. (1999). String theory and noncommutative geometry, *JHEP* **1999**, 09, p. 032.
Simões, T. J. M. (1980). *Análise da Formulação Funcional da Mecânica Quântica Não-Relativística*, Master's thesis, Universidade Federal do Rio Grande do Sul, Brazil.
Snyder, H. S. (1947). Quantized space-time, *Phys. Rev.* **71**, 1, pp. 38–41.
Sudarshan, E. G. C. and Mukunda, N. (1974). *Classical Mechanics: A Modern*

Perspective (John Wiley and Sons, New York).
Sundermeyer, K. (1982). *Constrained Dynamics* (Springer-Verlag, Berlin).
Symanzik, K. (1954). Über das Schwingersches funktional in der feldtheorie, *Zeits. Naturforsch.* **9a**, pp. 809–824.
Szabo, R. J. (2003). Quantum field theories in noncommutative spaces, *Phys. Rep.* **378**, 4, pp. 207–299.
Visconti, A. (1965). *Théorie quantique des champs* (Gauthier-Villars, Paris).
Weyl, H. (1950). *The Theory of Groups and Quantum Mechanics* (Dover, Princeton, N.J.).

Index